Miniature
Internal
Combustion
Engines

Miniature Internal Combustion Engines

MALCOLM STRIDE

THE CROWOOD PRESS

First published in 2007 by
The Crowood Press Ltd
Ramsbury, Marlborough
Wiltshire SN8 2HR

www.crowood.com

British Library Cataloguing-in-Publication Data
A catalogue record for this book is available from the British Library.

ISBN 978 1 86126 921 8

FRONTISPIECE: The first engine designed by the author, a 15cc
pushrod overhead-valve engine, shown fitted with a flywheel.

Typeset by Focus Publishing, Sevenoaks, Kent
Printed and bound in Singapore by Craft Print
International Ltd

Contents

Acknowledgements

It would be impossible to create a work of this nature without the help of many people, and I would like to acknowledge the assistance given by all those who provided support and encouragement during the writing of this book.

In particular, thanks must go to Rosina Tillyer for her support and assistance with proofreading the original document. Thanks are also due to Ingvar Dahlberg, Peter Gain, Ron Hankins, Brian Perkins, Alan Thatcher, Bill Conmore, Les Chenery and Mike Tull for the use of photographs of their models in the book.

I would especially like to thank all those who are interested in model I/C engines, because, without you, there would be no book.

Even a single cylinder engine has many individual parts. The 'kit' of parts for the author's 15cc single cylinder design seen assembled on page 30.

Introduction

BACKGROUND

Since the early days of model engineering, many model engineers have found internal combustion (I/C) engines a subject of great interest. Over the years there have been many notable contributors to the development of miniature engines. Indeed, some of their design concepts have proved to be influential in the development of full-size engines. Important contributors (among many others) have included Edgar Westbury, Gerald Smith, Elmer Wall, D.H. Chaddock, C.E. Bowden and, more recently, L.C. Mason, David Parker and Gordon Cornell, the latter concentrating on two-stroke engines.

Edgar Westbury in particular demonstrated that it was possible to develop small two-stroke and four-stroke engines capable of producing sufficient power for model boats and aircraft. Over many years he described his designs in his articles for *Model Engineer* magazine.

Many others contributed to the development process, mainly through necessity because commercially produced engines were not available until the late 1930s, when Walter Hurleman of the Junior Motors Corporation produced the Brown Junior two-stroke.

Even in the early days, several engine builders were producing working multi-cylinder and radial engine designs in spite of the limited facilities and materials available. Elmer Wall and Edgar Westbury produced several early four-cylinder designs, which, in Westbury's case, included the 15cc side valve *Seal* and the 30cc overhead-camshaft *Sealion*. Castings for both these engines are still available today. A notable series of radial engines was produced by Gerald Smith, including the three-cylinder *Osprey* and

The engine that brought I/C-powered models within the reach of many: the 1936 Brown Junior two-stroke. This is a 40th anniversary replica, produced in 1976.

Mastiff *flat-four four-stroke petrol engine to a design by L.C. Mason.*

five-cylinder *Buzzard* radials and an eighteen-cylinder double-row radial, which was an incredible achievement for the time (1924).

All of these model engineers showed what could be achieved and their achievements encouraged many others to have a go.

Recently, despite the ready availability of cheap commercial engines, there has been a revival of interest in the production of miniature I/C engines, with many more examples appearing at exhibitions and displays around the country.

THE PURPOSE OF THIS BOOK

It is the revival of interest in I/C engines that has prompted the writing of this book. The engines always arouse the curiosity of other model engineers, particularly at exhibitions, but even competent model engineers seem to hold the view that they could never build one themselves – a view usually expressed while admiring a complex multi-cylinder radial engine or similar. This book aims to show that, in fact, building an I/C engine requires no more skill than building other similar working models of engineering subjects.

This book is also aimed at those whose interest lies in internal combustion engines, who sometimes have the misconception that these require greater skill in construction than a steam engine, or who believe that model engineering is all about steam.

My hope is that this book will encourage everyone to 'have a go', no matter which group they fit into. I am not recommending that you start with a multi-cylinder rotary or radial engine but I would emphasize that building a single-cylinder four-stroke engine is within the scope of anyone who has produced a working stationary steam engine. Certainly, those who have built a steam locomotive will find building an I/C engine a straightforward exercise.

This book is intended to provide sufficient information about the processes involved in building miniature I/C engines, so that those with a basic knowledge of machining and workshop techniques will feel confident in tackling a simple single-cylinder engine design. Additionally, I hope to demonstrate that, although the processes may be different in some ways, they are no more complex than those involved in building a steam engine; both have a piston in a cylinder and valve gear to control the events in that cylinder.

Once a single-cylinder engine has been built, the jump to a multi-cylinder engine involves building on the skills acquired with the single. The main requirement is more patience in making similar components in larger numbers.

This book is not intended to be a detailed construction manual but will describe the techniques involved in producing the components required for an I/C engine. Where possible, photographs of actual set-ups have been used to illustrate the processes used.

I hope that those who have some experience of building I/C engines will also gain new information from this book, even if only a different approach to some of the techniques involved.

Above all, the book is meant to be enjoyed by all who read it, even those who have an interest in the subject but do not intend to build an engine. Hopefully, when you see engines at exhibitions in future you will have more appreciation of the techniques used by the builder.

ORGANIZATION OF THIS BOOK

Internal combustion engines come in a wide variety of types and configurations, ranging from simple single-cylinder two-stroke engines with three moving parts, through to multi-cylinder radial and rotary four-stroke engines with large numbers of components.

To avoid duplication, this book is split into three parts:

Part 1 (chapters 1–3) covers the basic operating principles of the various types of engine followed by comments on basic design aspects, together with a discussion of the workshop equipment needed.

Part II (chapters 4–18) provides a detailed description of the processes and techniques involved in producing a single-cylinder four-stroke engine. Most of this section is applicable to engines whether they are single- or multi-cylinder.

Part III (chapters 19–21) provides the additional information needed to produce two-stroke and multi-cylinder engines of various types, including radial and rotary engines.

Two-stroke engine construction is described in a dedicated chapter, but, again, where the techniques described for four-stroke engines apply, they are not repeated. Readers who have no experience of internal combustion engines should therefore read the earlier chapters even if they intend to build a two-stroke or a multi-cylinder engine.

Part I: The Basics

1 Engine Types and Operating Cycles

WHAT IS AN INTERNAL COMBUSTION ENGINE?

An internal combustion (I/C) engine, as the name implies, is one in which the fuel is burnt inside the engine cylinder to produce the power. In contrast, a steam engine is an external combustion engine, where the fuel is burnt outside the engine, heating water in a boiler which produces the steam that drives the engine. Since the fuel is burnt inside the engine, an I/C engine is more efficient than a steam engine because the overall heat losses are much lower.

The fact that the fuel is burnt inside the cylinder raises one significant problem: how is it ignited? Many different methods have been tried over the history of I/C engines, including hot bulbs, spark ignition, glow-plug ignition and compression ignition/diesel. Nowadays the most commonly used options are spark ignition, diesel and, for model engines in particular, glow-plug ignition (see the relevant chapters for each engine).

The other issue with I/C engines is how to get the fuel/air mixture into the engine. Because a steam engine runs at relatively slow speeds (generally around 300rpm maximum) the valve gear does not need to operate at high speed. Internal combustion engines, on the other hand, run at speeds of up to several thousand revolutions per minute. Because the valve gear has to operate efficiently at this higher speed, its method of operation is significantly different from that of steam engines.

There are two main types of I/C engine in common use: the two-stroke engine and the four-stroke engine. The operating cycles of these two ypes are very different and builders of either type of engine need to have a good understanding of them.

OPPOSITE: Pushrod single-cylinder air-cooled four-stroke engine, by the author.

LEFT: Typical air-cooled crankshaft-induction single-cylinder two-stroke engine.

TWO-STROKE ENGINES

Some History

The person credited with the invention of the ported two-stroke engine is Sir Dugald Clark, who developed the first such engine in 1867. The design was simplified and patented by Joseph Day from Bath in the United Kingdom in 1891 and the two-stroke cycle commonly used is often known as the Day Cycle.

The big advantage of the two-stroke engine is its inherent simplicity. A single-cylinder two-stroke engine has only three moving parts: the crankshaft, the piston and the connecting rod. In this basic form the intake and induction events are controlled by the piston uncovering and covering ports cut into the cylinder wall.

An important feature of the two-stroke cycle is the use of the crankcase as a pump to propel the fuel/air mixture into the cylinder. This also means that the crankcase has to be sealed and that lubrication is by means of oil carried into the engine with the fuel.

The Two-Stroke Cycle

As the name suggests; all the operating events in a two-stroke take place during two complete strokes of the piston or one revolution of the crankshaft. The diagram opposite shows the two-strokes and the events that take place during their operation.

The Power Stroke

In the first part of the diagram [opposite], the engine has just fired and the piston is starting to descend on the power stroke. At the same time, the fuel/air mixture in the crankcase is being compressed.

Further into that stroke, the exhaust port and transfer port are open and the burnt mixture is exiting the cylinder, with the fresh mixture entering via the transfer port and being turned towards the cylinder head by the deflector on the piston. This assists the removal (scavenging) of the burnt gases from the cylinder.

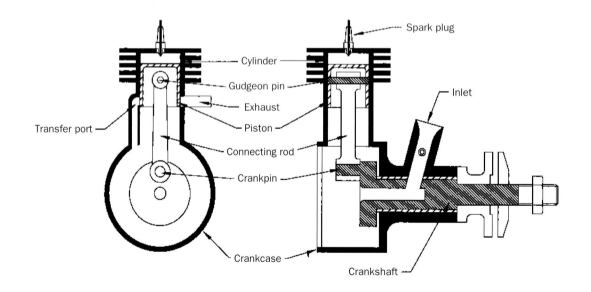

Diagrammatic cross-sectional view of a typical crankshaft-induction two-stroke engine.

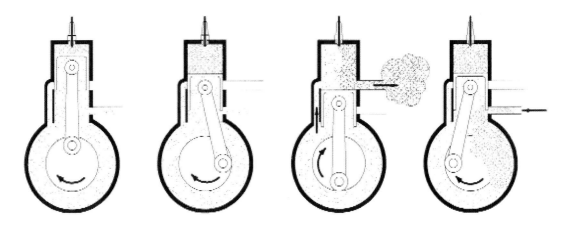

The basic two-stroke cycle: (left to right) the point of ignition; the end of the power part of the cycle; the exhaust open and transfer of new mixture; and finally compression of the mixture and crankcase induction.

The Compression Stroke

As the piston goes past bottom dead centre and commences the upward stroke, the transfer port is closed and the mixture is being compressed in the cylinder. Further up that stroke, the inlet port opens and fresh mixture is drawn through the carburettor into the crankcase (induction).

At top dead centre, the compressed mixture in the cylinder is ignited and the cycle starts all over again.

These events are often illustrated using a timing diagram based on crankshaft rotation angle.

Timing

In this basic type of two-stroke engine, the events are symmetrical about top dead centre and such an engine is capable of starting and running in either direction. This is particularly true of diesel and glow-ignition types in model form. Spark-ignition engines are biased towards one direction of rotation, because the spark is set to fire just before top dead centre (in other words, 'advanced'), and in order to run backwards the ignition timing must be altered.

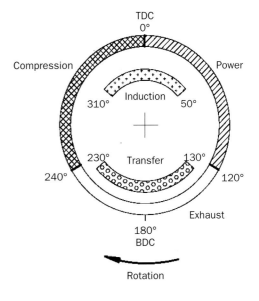

Typical two-stroke engine timing diagram.

Alternative Inlet Options

This basic engine design does not allow for very high performance and, in order to improve performance, different inlet porting arrangements have been developed.

Sectioned two-stroke glow-plug engine, showing the internal components and in particular the crankshaft-induction layout.

One very common method involves controlling the induction timing by means of a cut-out in the crankshaft leading into a passage through the centre of the shaft, and thence into the crankcase.

Another often used option is the use of a rotary disc or drum valve at the rear of the crankcase driven by an extension of the crank pin. This gives greater design freedom than the crankshaft port because, even with long induction periods, there is no danger of weakening the crankshaft.

The use of either of these options means that the timing of the inlet events is not connected to the movement of the piston, with the result that the inlet timing can be set solely on the basis of obtaining optimum performance.

Another option found on many of the well-known commercial Cox engines is the use of a reed valve, which is lifted by the reduction in pressure in the crankcase. This is caused by the piston rising, thus allowing the mixture to enter. When the piston goes past top dead centre on to the down stroke, the pressure increases. The reed goes back on to its seating, sealing the crankcase, allowing the mixture to be compressed and pushed into the cylinder via the transfer port. The reed system can be a very good option for multi-cylinder two-stroke engines.

Scavenging

One of the problems with the two-stroke cycle is that the exhaust port and transfer port are both open at the same time for part of the stroke. This means that, if measures are not taken to prevent it, the fresh mixture entering via the transfer port can travel straight across the cylinder and out through the exhaust port, thus wasting fuel. Also, exhaust gases may remain at the top of the cylinder, preventing a full charge of fresh mixture getting into the cylinder.

This incomplete scavenging of the cylinder is dealt with in a number of ways. The simplest is to have a vertical deflector on the top of the piston, to deflect the incoming fresh mixture upwards to the top of the cylinder, thus replacing the burnt gases with fresh mixture and also stopping the mixture travelling straight across the cylinder. The problem with this option is that, in order to keep the compression ratio high enough, the deflector may have to go into a slot in the cylinder head at the top of the stroke. This obviously complicates the layout of the cylinder head and makes both it and the piston more difficult to make. It is also not generally a practical option for compression-ignition engines.

Another option is to angle the cylinder liner transfer ports upwards so that the incoming mixture is again directed towards the top of the cylinder. This is known as 'loop scavenging' and is used on many high-performance engines in various forms; one of the most common is 'Schnuerle' porting, in which two transfer ports are arranged opposite each other. This arrangement also incorporates an extra 'boost' port (opposite the exhaust port), which opens slightly later than the transfer port and directs an extra boost of fresh mixture upwards to the top of the cylinder, further assisting the scavenge process.

Various other porting arrangements have been used in order to overcome the scavenging problem. These include reverse flow, in which there are often three exhaust ports and three transfer ports with the inlet flow directed up towards the top centre of the cylinder; and split flow, which has two pairs of ports, again with the inlet flow directed to the top of the cylinder.

The aim of all these arrangements is to ensure that the incoming mixture is directed in such a way as to assist the removal of exhaust gases and ensure that the cylinder is fully charged with fresh mixture for each two-stroke cycle.

Uniflow Porting

Another porting arrangement sometimes used is the 'uniflow' system, in which the transfer ports are arranged normally but the exhaust port is located in the cylinder head. It may take the form of a poppet, sleeve or disc valve. This arrangement means that the incoming fresh mixture does not change direction and flows straight from the bottom to the top of the

A high-performance Hornet *rear disc-induction two-stroke engine.*

cylinder. The disadvantage of this arrangement is that it complicates the engine because of the need to have some form of separate drive arrangement for the exhaust valve.

Most modern two-stroke designs seem to have some form of crankshaft or rear disc-controlled induction, with a loop scavenge cylinder porting arrangement.

Multi-Cylinder Two-stroke Engines

Two-stroke engines have been made in a variety of multi-cylinder layouts including in-line, 'V', flat and, occasionally, radial arrangements. There have also been many attempts to use a second cylinder as a replacement for the crankcase to pump the fresh mixture into the cylinder. This arrangement is known as the 'split single'.

Induction Options with Multiple Cylinders

One of the problems with the multi-cylinder two-stroke engine relates to the use of the crankcase as the means of getting fresh fuel/air

mixture into the cylinder. Because of this basic requirement, each section of the crankcase has to be sealed from the next in order that the crankcase pressure under each cylinder is dependent only on the movement of that piston, and is not affected by any of the other cylinders.

This limits the type of induction control that can be used and for anything other than two cylinders; this means the use of side port induction or separate reed valves in the crankcase side, one for each cylinder.

Induction with In-Line Twin Engines

For in-line two-cylinder engines there are several ways of using other induction options. If the crankshaft is arranged so that both cylinders are in phase – in other words, both pistons travel up and down the cylinder in parallel with each other – then any of the induction options described for the single-cylinder engine can be used. In this situation, the crankcase pressure behaves in the same way as a single-cylinder engine, although, during the

SOP twin-cylinder two-stroke glow-plug engine.

transfer part of the cycle, the pressure is being used to feed two cylinders at the same time.

This arrangement does not seem to give any great advantage over a larger single-cylinder engine because there is only one firing stroke and the arrangement does not provide any benefits in terms of improved balance. In fact, because of the complexity of the crankshaft and the probable need for split big-end bearings, it does seem to have some disadvantages. Despite this, the different dimensions (particularly the reduced height) of the twin-cylinder engine may be advantageous in some situations.

Another option with in-line twin-cylinder engines is to use a disc-induction system between the cylinders, making the engine effectively two single-cylinder engines bolted together. The crankshaft is in two halves, with the front part similar to a normal single-cylinder crankshaft. This drives the induction disc for that cylinder and is connected via a length of shaft in a plain bearing to the second disc, which links to the crank pin for the second cylinder. The plain bearing between the cylinders forms a seal so that the two parts of the crankcase are isolated from each other. This engine has the advantage of good balance since the pistons are 180 degrees out of phase, with one balancing out the other and also because the cylinders fire alternately; there are two firing strokes per revolution and this provides smooth power. The best-known engine with this layout is probably the *Taplin Twin*, which was, in fact, developed using two E.D. Comp Special cylinders mounted on a new crankcase.

It is also possible to use 'crankshaft' induction in the same way by having identical halves of the crankshaft connected via a similar length of plain bearing but with separate induction passages at each end of the engine. This type of engine is effectively two single-cylinder engines bolted back to back, with a short length of shaft connecting the two crankshafts together. This method of construction has been used to produce two-cylinder engines in the past by bolting two commercial engines back to back.

One advantage of the arrangement is that there are two exposed ends to the crankshaft that can be used for the drive. In a boat, for example, the flywheel could be mounted on one end and the drive to the propeller taken from the other end. This arrangement also makes starting with a cord round the flywheel much easier. In a model racing car, the engine could lie across the body, with one wheel mounted on each end of the shaft, making a very simple direct-drive arrangement.

Induction with Flat-Twin Two-stroke Engines

Flat-twin engines suffer no such problems because, if the crank is arranged so that each cylinder reaches the top of its stroke at the same time, the system will operate in exactly the same way as a single-cylinder engine. In this case, both pistons will be compressing the mixture in the crankcase at the same time, thus feeding both cylinders. This type of engine is very well balanced because the weights of the reciprocating parts cancel one another out and both cylinders fire at the same time. Because the 'push' is in opposite directions, the impulses cancel each other out.

This type of engine is also fairly easy to make, with only the crankshaft being more complex than that of the single-cylinder engine.

Flat engines with more than two cylinders suffer similar restrictions to multi-cylinder in-line engines. In this case the crankcase needs to be sealed only between each pair of opposed cylinders if the cranks are arranged so that adjacent cylinders form a balanced pair (a flat-twin). This makes a six-cylinder flat-twin easier to make than a six-cylinder in-line engine, which must inevitably have side-port or reed induction and has each cylinder sealed from its neighbours.

V-Format Engines

Two-stroke 'V' engines are not really a practical option. Normally each of the cylinders of a pair in a 'V' arrangement operates on the same crank pin, making it impossible to isolate the

A very unusual 60cc six-cylinder water-cooled two-stroke engine by Eldon.

cylinders from each other. This means that the pumping action is dependent on two pistons that are on different parts of the cycle (typically 90 degrees apart), with each at a different point of their stroke at any given time. This reduces the effectiveness of the crankcase pumping action; the induction timing (into the crankcase) must be set to suit the action of both cylinders and is therefore a compromise. This drastically reduces the available induction and transfer periods, making such a layout impractical without the use of some sort of additional crankcase pumping cylinder.

There have been some full-size V-twin two-strokes but these have used such arrangements as separate crankshafts linked to a common drive shaft or separate crank pins spaced apart at an angle based on the angle of the 'V'. Other arrangements include using separate pumping cylinders to pump the mixture into the power cylinders, as in the Funck six-cylinder rotary two-stroke design of the early 1900s.

FOUR-STROKE ENGINES

Development of the Four-Stroke Cycle

The first practical four-stroke internal combustion engine is credited to Jen Lenoir in 1858. It was unusual, both because it used coal gas as fuel and also because it was a double-acting engine with a power stroke acting on both sides of the piston.

Nikolaus Otto built the first really practical four-stroke internal combustion engine in 1867, and the design was later improved upon by two ex-employees of his, Gottlieb Daimler and Wilhelm Maybach. Daimler incorporated the carburettor, making the engine capable of using liquid fuels, whilst Maybach constructed the first four-cylinder engine in 1890. Otto's most notable claim to fame is that he patented his design thus preventing anyone else from using his four-stroke cycle, which became known as the Otto Cycle (*see* page 20). This patent restriction led to many attempts to develop other engine cycles in order to attempt to circumvent the patents.

The patent was eventually invalidated in 1886 when it was found that Alphonse Beau de Rochas had described the four-stroke principle in a private paper some time before Otto raised his patent.

Difference Between Four-Stroke and Two-Stroke Engines

The most noticeable difference between a four-stroke engine and a two-stroke engine relates to the number of moving parts. In a four-stroke engine the induction and exhaust events are controlled by mechanical valves driven from the engine crankshaft, normally through a 2:1 reduction gear arrangement driving a camshaft. Cams on this shaft operate the valves, either directly or, as in earlier engines, via pushrods. In

Single-cylinder engine showing the pushrod-operated valves.

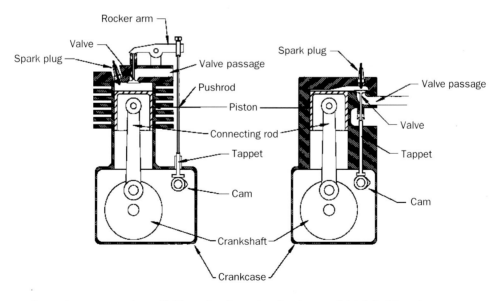

Four-stroke engine cross-sections: (left) pushrod overhead valves and (right) side valves.

current designs the valves are usually in the cylinder head (overhead valves) although many earlier designs used a side-valve arrangement.

The additional parts on four-stroke engines make them heavier than comparable two-strokes. However, because the exhaust generally opens later and there are half the number of firing strokes, they run more quietly. In addition, they offer the significant advantage in today's world of producing far less pollution, partly because lubrication is normally by means of engine-driven pumps providing pressure lubrication to all the moving parts, so that oil no longer needs to be incorporated into the fuel.

The Otto Four-Stroke Cycle

The Otto Cycle takes place over four strokes of the piston, which means that the engine fires only once every other revolution of the crankshaft. The advantage of the four-stroke engine is that the valve events are dependent only on the timing of the cams which means that the timing of the opening and closing of inlet and exhaust can be set to provide the most desirable operating parameters for any given design of engine. There are four-stroke engines designed to give minimum fuel consumption, absolute maximum power, high torque, high rpm or minimum pollution.

The four strokes of the Otto Cycle are the following (see the illustration below):

1. the inlet stroke;
2. the compression stroke;
3. the power stroke; and
4. the exhaust stroke.

The illustration shows a cross-section of a single-cylinder overhead-valve four-stroke engine, clarifying the valve positions during the four strokes. The valves are operated by pushrods from the cams, and are shown on either side of the engine for clarity; they would normally both be operated by one camshaft at the side of the engine.

Inlet Stroke
The first part of the diagram shows the inlet stroke: the inlet valve has just opened, the

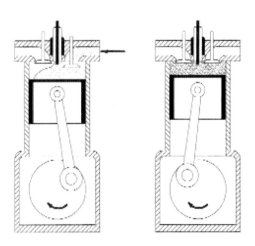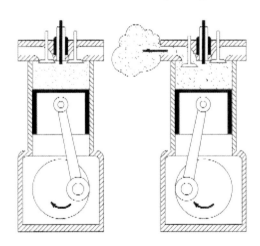

The four strokes of the Otto Cycle: (left to right) inlet stroke, with the inlet valve open and fuel/air mixture entering the cylinder; compression stroke, with the valves closed and the mixture being compressed by the rising piston; after ignition at top dead centre, power stroke, with the burning mixture pushing the piston down; finally, exhaust stroke, with the rising piston expelling the burnt gases out of the open exhaust valve.

exhaust has just closed and the piston is beginning to descend. This draws fuel/air mixture from the carburettor in through the inlet valve to fill the cylinder.

Compression Stroke
The second part of the diagram shows the piston beginning to rise in the cylinder with the inlet valve having just closed, so the rising piston is compressing the fuel mixture in the cylinder.

Power Stroke
As the piston reaches the top of the stroke the mixture is ignited and the resulting expansion of the burning gases pushes the piston down on the power stroke.

Exhaust Stroke
At the bottom of this stroke, the exhaust valve opens and the rising piston then pushes the burnt gases out of the cylinder via the open exhaust valve. At the top of this stroke, the cycle commences again from the start.

The total cycle is often colloquially described as 'suck, squeeze, bang, blow'.

Timing Considerations
One important point to make is that, in practice, the valves do not open or close exactly on the top or bottom dead-centre points during the cycle. This is in order to compensate for the effects of inertia on the gas flows. For example, when the inlet valve opens, the column of fresh mixture in the inlet tract takes a finite time to accelerate into the cylinder. In a typical four-stroke engine, the inlet valve will open some time before top dead centre and will close after bottom dead centre, whilst the exhaust valve will open before bottom dead centre and close after top dead centre. The valves are therefore open for longer than the 180 degrees of a stroke. This means that, at times during the cycle, both valves are open together (*see* timing diagram overleaf).

Expressing Timing
The timing of the valves is normally expressed in terms of the relationship of their opening and closing points to the top or bottom dead-centre positions and is measured in terms of degrees of crankshaft rotation. For the engine shown in the (*see* timing diagram overleaf), the exhaust would be said to 'open 34 degrees before BDC and close 33 degrees after TDC'. This open period before the relevant dead centre is often called the 'inertia period'.

Timing Diagrams
Because the four-stroke cycle occupies two revolutions of the crankshaft, there are two ways of drawing the timing diagrams for four-stroke engines (see page 39).

The first shows each revolution as one half of the circle so that each stroke occupies one quarter-circle. This type of diagram is easier to understand for those new to such things because the complete cycle is laid out.

The other way of drawing timing diagrams shows the two revolutions superimposed on each other; the power and compression strokes are less obvious, although they are still there.

Valves and Valve Gear
The majority of four-stroke engines use the cam-operated poppet valve because of the ease of construction and its simplicity. Even with poppet valves, however, there can still be significant variation in engine layout and the method of operation of the valves.

Some of the more common options include the side-valve layout, pushrod-operated overhead valves, and overhead camshaft, where the cams operate directly on tappets located on ends of the valve stems. The latter is by far the most common arrangement on modern engines because it keeps the weight of the moving parts of the valve gear to a minimum, thus avoiding problems due to inertia.

It must not be forgotten that in a four-stroke engine running at 6000rpm, each valve is opening and closing fifty times a second. At this speed, heavy valve gear will increase the

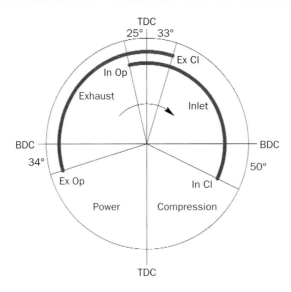

A timing diagram showing all four strokes, one in each quarter circle. The diagram shows two complete revolutions of the crankshaft and is expressed in terms of camshaft angle. The other type of diagram (see page 39) relates to crankshaft angle.

load on the cams and may cause the valve closure to lag behind the cam position, thus affecting the engine timing and performance, and also potentially damaging the valve gear. In the past, a hybrid pushrod layout was tried, with the camshaft put higher up in the engine; in this arrangement, the camshaft could still be driven by spur gears but with shorter, and therefore lighter, pushrods. The idea was to reduce the weight of the valve gear without using chain or shaft drives to the camshaft.

The basic layout described above is not the only method of producing a four-stroke cycle. The cam-operated poppet valve has in the past

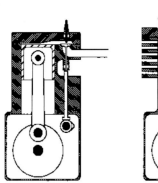 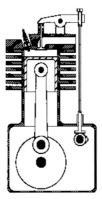 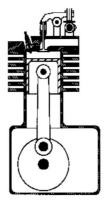 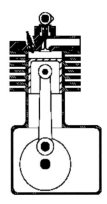

Four-stroke valve layouts: (left to right) pushrod side valve, pushrod overhead valve, overhead cam with rocker arm, direct-acting overhead cam. The camshaft drives are omitted (for clarity), but will be by gear, chain or belt from the crankshaft.

An Edgar Westbury design, the Wyvern *gas engine, built by the author.*

been replaced with different types such as sleeve or disc valves, with consequent variations in engine layout.

Because the four-stroke cycle does not rely on a sealed crankcase to pump the mixture into the cylinder, many different arrangements of cylinders are possible. In fact, the earliest four-stroke engines owed much to steam engine design, with open cranks, long strokes and a layout very like a stationary steam engine. Some of these engine designs did not use a mechanically operated inlet valve. Instead, the valve was held closed by a light spring and was pulled open by the suction created when the piston descended on the inlet stroke. This arrangement was not very efficient and was soon dropped.

Timing for Different Engine Types

The four-stroke engine using cam-operated valves has no real restrictions on the valve timing set-up, so it can be set up to provide the characteristics most desirable for the particular application. Typical differences in timing can be seen by comparing the timing diagram of two very different designs by Edgar Westbury: the *Wyvern*, a typical horizontal gas engine, and the *Sealion*, a high-performance overhead-valve engine.

Because the gas engine is designed to run at low rpm, and typically at a steady speed, the inertia periods are quite short, with the inlet opening only 5 degrees prior to top dead centre and closing 45 degrees after bottom dead centre. The exhaust opens 55 degrees before bottom dead centre and closes 15 degrees after top dead centre.

In contrast, the *Sealion* overhead-camshaft, four-cylinder engine is designed to run at high speed. The inlet opens 20 degrees before top

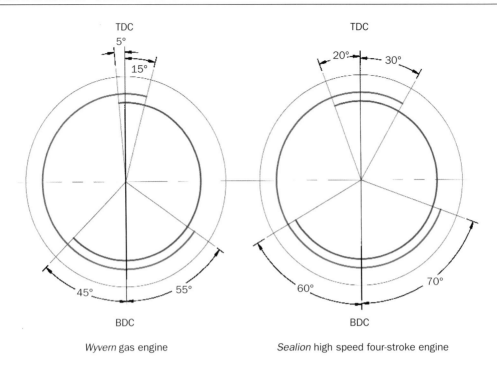

TDC
5°
15°
45°
55°
BDC

Wyvern gas engine

TDC
20°
30°
60°
70°
BDC

Sealion high speed four-stroke engine

Gas engine and Sealion *high-performance engine timing comparison, showing the much more advanced timing of the high-revving* Sealion *compared with the slow-revving* Wyvern *gas engine.*

dead centre and closes 60 degrees after bottom dead centre, with the exhaust opening 70 degrees before bottom dead centre and closing 30 degrees after top dead centre.

Other factors affected by the valve timing are the ease of starting and the speed at which an engine will tick over reliably.

Multi-Cylinder Four-Stroke Engines

Unlike two-stroke engines, four-stroke engines may be constructed in many different multi-cylinder configurations, the only restriction being the ability to machine the parts and to be able to fit every thing together. Even in model sizes, eight-, ten- or twelve-cylinder engines are now becoming common. Often the main limiting factor is the ability to provide a reliable ignition system; this is often circumvented by using glow-plug ignition, which does not require the complexity of a distributor and coil(s).

Lubrication is also a consideration but this can be simplified by the use of needle rollers or ball races for the bearings. Such bearings do not require a pumped oil supply and, as a result, 'splash' lubrication is once again becoming quite common. This avoids the need to drill long small-bore oil-ways through components such as the crankshaft, making construction much easier.

Construction of camshafts is now also made easier with the use of modern adhesives such as those provided by Loctite.

Often the real factor in deciding to build such engines is the patience of the builder: a V12 has twenty-four valves, twenty-four cams, twelve cylinders, and so on.

One of the features of all multi-cylinder engines is that the cylinders are designed to fire in a particular order to give the smoothest power output and lowest vibration levels. In this respect, some engine layouts are better than others. A six-cylinder in-line engine ('straight six') is generally a very well-balanced engine because the crank pins are evenly spaced at 120-degree intervals.

Radial and Rotary Engines

Radial and rotary engines were first developed in the early days of aviation. They were able to produce a high power output for a given size and, most importantly, because the cylinders did not shield each other as in an in-line engine, air cooling was effective. They have a special niche in the world of miniature engines. They look and sound impressive in operation and many have been constructed over the years. The two types of engines are very similar – in both cases, the cylinders are arranged 'radially' around the central crankcase, with the cylinder heads facing outwards from the centre.

The radial engine still finds favour today where there is a need for a light, compact power plant; the rotary engine is rarely found other than in preserved or miniature versions.

The difference between the two types is implied in the names. In a radial engine, the main body of the engine is fixed to the aircraft and the crankshaft rotates within the engine as normal. In a rotary engine, the crankshaft is fixed and the whole engine rotates around it. The propeller is mounted on the engine and the crankshaft is fixed to the aircraft.

The rotary engine has several unique design features, particularly the need to take in the fuel/air mixture via the crankshaft. The fuel/air mixture then travels through the crankcase before reaching the valves and cylinders.

Radial/Rotary Operating Cycle

Radial and rotary engines use the standard four-stroke cycle but there are certain peculiarities in the way they operate. Both types are normally designed so that the cylinders fire alternately around the circle. Thus, the firing order for a seven-cylinder engine will be 1, 3, 5, 7, 2, 4, 6. This is said to provide smoother operation than would be achieved if the cylinders fired in sequence.

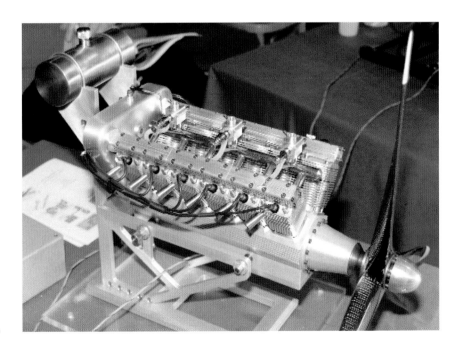

A V12 glow-plug engine, designed and built by Ron Hankins, seen here at the Model Engineer Exhibition.

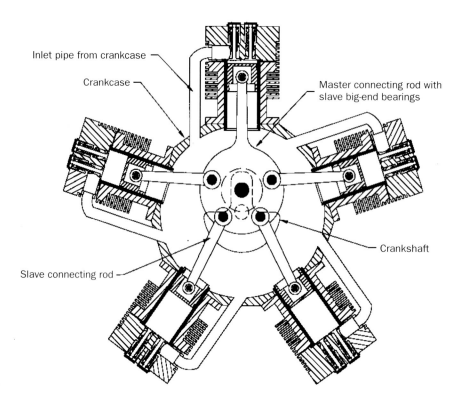

Inlet pipe from crankcase

Crankcase

Master connecting rod with
slave big-end bearings

Crankshaft

Slave connecting rod

ABOVE: *Radial engine layout, showing the cylinder arrangement and master/slave connecting-rod assembly. The valve gear is omitted for clarity.*

LEFT: *A part-assembled Bristol Aquila sleeve-valve radial engine, showing the general layout and the gear drives to the sleeves.*

Characteristic Features

One characteristic feature of rotary engines is the fixing of the carburettor to the end of the crankshaft and the way the fuel mixture is drawn into the crankcase before entering the cylinder. This is necessary because the crankcase and cylinders rotate, whilst the crankshaft is stationary.

Radial engines often adopt a similar arrangement with the carburettor mounted on the rear of the crankcase. Some engines have been built with the carburettor feeding into a circular inlet manifold joining all the inlet valves.

Lubrication of radial engines is normally by means of oil fed into the crankcase by means of a pump or gravity and then pumped back to the external oil tank. In a rotary engine, the oil is carried round the engine with the fuel, as in a two-stroke, and the 'used' oil is ejected from the exhaust.

The operation of the valves is usually by pushrods but the cam used is often of a different type from that employed on a normal four-stroke. Many radial and rotary engines use a cam ring that is normally driven in the opposite direction to the engine rotation and rotates at a fraction of the engine speed. For example, in one nine-cylinder rotary engine for which I have drawings, the cam rotates at one-eighth of the engine speed and has four exhaust and four inlet lobes. This arrangement provides the alternate firing order and the correct valve operation for each cylinder.

Some engines are built with individual gear-driven camshafts for each cylinder. With this arrangement, there can be a problem fitting the camshaft drive gears into the crankcase with the larger cylinder numbers. Many full-size radial engines used sleeve valves but this method of construction provides its own special problems (*see* Chapter 20).

Rotary/Radial Engine Big-End Design

Special arrangements are necessary for the big end of rotary and radial engines because there is more than one connecting rod to connect to the single crankpin.

The most common arrangement is to have

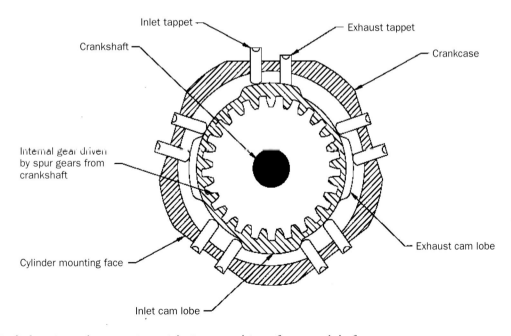

Radial engine valve operation with ring cam driven from crankshaft.

*Example of a master/
slave radial engine
big-end layout.*

one master connecting rod with a normal big end and then to link all the other (secondary or slave) connecting rods to this, with their big ends positioned in a circle around the master big end. This does have some disadvantages due to the geometry involved (*see* page 155).

The Manley Baltzer big end used in some engines has one master connecting rod but the slave connecting rods have a circular quadrant in place of the normal big-end bearing. These quadrants are held in close-fitting circular grooves machined in the two halves of the master big end. This arrangement ensures that each connecting rod is in the correct position and has the same effective length.

Single and Double Row Engines

The requirement for ever-increasing amounts of power from engines has led to an increasing number of cylinders being used. It soon became apparent that it was not practical to accommodate more than nine cylinders around one crankshaft, and this led to the development of 'multi-row' engines, with more than one circle of cylinders. Typical configurations were eighteen- and fourteen-cylinder double-row engines, although triple-row engines with up to twenty-seven cylinders were also built.

These configurations obviously need a different crankshaft but the basic layout is the same as for a single-row engine. The usual way of ensuring adequate cooling for all cylinders is to stagger the cylinders to avoid the rear cylinders being shielded from the cooling airflow by the front row.

Normally, single-row engines have an odd number of cylinders and so a double-row engine will have an even number. Some miniature single-row engines have been built, with even numbers of cylinders. They usually use conventional gear-driven cams to provide the correct timing and firing order.

2   Engine Design Basics

INTRODUCTION

When constructing an internal combustion engine of any type, it helps to have an appreciation of the design principles and constraints involved. This will give the builder an understanding of why certain things have to be done in a particular way, and also an appreciation of which items are critical to the successful building of an engine. Although this book is aimed at those new to I/C engine building, it is hoped that readers may be encouraged to progress further in the field and perhaps think about designing their own engine. Any experienced model engineer is capable of designing and building a successful single-cylinder engine, given some basic under-standing of the principles involved. The lessons learned from such an exercise can then be applied to more complex multi-cylinder engines.

To illustrate the design process, the text will describe that which I have used when designing engines, but will comment on other options where relevant. I will be describing the design of a single-cylinder overhead-valve four-stroke engine and will go on to comment on some of the different approaches to two-strokes and multi-cylinder engines.

At this stage, it is important to emphasize two points:

- a CAD program will be of great use during the design process, allowing even significant changes to the engine layout to be made with ease. It will also help to ensure that the different components will fit together correctly;
- a great deal of information can be gleaned by examining existing successful designs, to gain an appreciation of the basic proportions for which to aim (*see* page 45). It is worth remembering, though, that commer-cial designs may use exotic or especially heat-treated materials, and any light-weight, high-performance design should be approached with a degree of scepticism when considering your own design.

STARTING OUT

What Not To Start With

It is often useful to eliminate some possibilities before starting, to avoid going down a blind alley at a later stage. When designing a first engine, it is sensible to avoid such things as four-valve cylinder heads, hemispherical heads with angled valves, and multi-cylinder engines.

Where to Start?

First, set out some basic parameters for the engine in question:

- operating cycle (two- or four-stroke);
- cylinder capacity;
- bore and stroke;
- fuel to be used;
- compression ratio;
- type of valves and method of operation;
- ignition type; and
- performance required.

A 15cc overhead camshaft engine designed by the author.

The latter will influence some of the other parameters and so should be considered at a very early stage. Other options to be decided at this stage will include the method of cooling, lubrication arrangements, and carburettor type.

Finally, one of the most significant things to take into account at this stage is the equipment available for machining the engine. After all, if you have a 3½in lathe with a vertical slide and no other milling equipment, you will need to pick a size of engine that can be machined easily using those facilities.

First Decisions

Let us assume that you want to design a single-cylinder four-stroke engine: most builders prefer the four-stroke engine, and many of the decisions that need to be taken on a four-stroke engine are also applicable to a two-stroke engine.

At the very beginning, you need to make a decision on cylinder capacity and the type of performance required. The reasoning behind this is that cylinder capacity is probably the most basic parameter on any engine, while the type of performance required will have an impact on other parameters, which themselves have a relationship to the cylinder capacity.

For example, in the case of a medium-performance engine with a cylinder capacity of around 15cc, aiming for medium performance implies a bore and stroke that are approximately equal, or with a slightly longer stroke compared to the cylinder bore. In other words, you will need a square or under-square bore to stroke ratio.

A bore of 25mm and a stroke of 30mm gives a swept volume of approximately 14.7cc, which is close enough to the desired nominal 15cc capacity.

BASIC ENGINE LAYOUT

Having set the bore and stroke, it is now easy to start to define the basic layout of the engine. Because the stroke has been fixed, the crank throw is set and you will be able to create a rough layout with an outline crankshaft and cylinder bore in their correct relationship. This will enable you to set the piston outline

dimensions and, from this, the connecting rod length.

The piston is typically made with the length equal to the bore, or slightly longer for this type of engine. The centre of the small-end bearing should be set about halfway down the piston. This will fix the minimum length of the connecting rod, because at the bottom dead centre position the piston skirt must be clear of the crank web.

At this stage, it is helpful to have a trial run and put the basic piston and connecting rod outlines in their correct relationship. A check should be made on the basic clearances between the connecting rod and the bottom edge of the cylinder when the crankpin is near the 90-degree point with the connecting rod at its most angled position. When using a CAD program (*see* page 45), this is typically done by creating the piston and connecting rod as objects that can then be placed in the correct positions.

If the layout is being produced on paper, cut out paper outlines of the piston and connecting rod with the basic centres marked and check the layout by hand. (NB: with paper drawings you

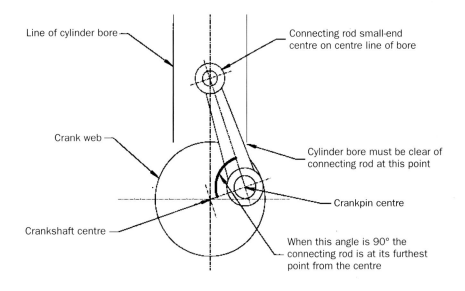

Line of cylinder bore

Connecting rod small-end centre on centre line of bore

Crank web

Cylinder bore must be clear of connecting rod at this point

Crankpin centre

Crankshaft centre

When this angle is 90° the connecting rod is at its furthest point from the centre

Checking the minimum clearance between the connecting rod and cylinder using a CAD program.

will need to produce a set of development drawings prior to a full set of working drawings. With CAD programs, the drawings can gradually evolve from one to the other as details get firmed up.)

All that is needed is to move the centre of the connecting rod small end up and down the cylinder centre line, whilst the big-end centre is moved around the crankpin rotation circle. The highest point at which the edge of the connecting rod intersects the lines of the bore extended downwards is the lowest point at which the bottom of the cylinder can be set. However, do not forget that at this point the piston lower edge may be close to the cylinder lower edge, meaning that you need to allow for the thickness of the piston skirt when checking clearances.

After these checks, the basic layout will have the connecting rod centres fixed, the cylinder lower end position and bore fixed, and the piston stroke and hence the crankshaft throw fixed.

FURTHER DECISIONS

In order to progress the design process, further decisions – relating to the valves and how they are to be driven, the ignition system, and how the crankshaft will be supported – need to be made.

Valves

Is the engine to be side-valve, overhead-valve or sleeve-valve?

For more on sleeve-valve engines, *see* Chapter 20; these valves raise their own peculiar set of issues and are not suitable for a first attempt at building. Angled valves should also be avoided on a first design because of the increased difficulty of constructing the cylinder head and piston.

For a medium-performance engine, over-head valves are probably the most sensible option; they are easier to set out, because side valves and their operating gear have to fit into a more confined space.

You now need to decide on the number and size of valves. In model engines, it is a good idea to fit the largest size of valve possible. This size is governed by the number of valves and the cylinder bore.

In a first engine, it is sensible to adhere to two valves per cylinder, which means that the maximum size of valve head that can be incorporated into a cylinder head is approximately one-third of the bore size. This allows for the spark or glow plug and clearance between the two valves. Do not forget that it is not only the inside of the cylinder to be considered but also fitting in the tappets and rockers, and so on, on the outside.

Piston and con-rod assembly from a 15cc four-stroke engine.

Valve and spring assembly, showing an E-clip retainer for the spring.

Built-up camshaft from a single-cylinder engine.

Camshafts

How will the valves be operated? The typical four-stroke engine has poppet valves operated by an engine driven camshaft running at half crankshaft speed. Cams on this shaft move tappets, which either act directly on the valves, as in the case of some overhead-camshaft engines, or on rocker arms that transmit the movement to the valve stems (in the case of other overhead-camshaft or pushrod engines).

The camshaft is driven at half engine speed by gears, chains or, in more modern engines, flexible, toothed belt drives.

Overhead-camshaft engines may have the camshaft in two locations:

1. immediately above the valves, with the cams operating the valves via tappets acting directly on the ends of the valve stem. Adjustment, if provided, is by means of shims between the two;
2. to one side of the valves, with the cams operating the valves via rocker arms running on pivots between the camshaft and valves. Adjustment is usually by means of screws and locknuts in the rocker arms. (The problem with this arrangement is that it provides for a very crowded cylinder head and it may be necessary to use bent rockers, as on the Edgar Westbury *Sealion* design.)

Pushrod engines can use a gear or toothed belt drive, but typically two or three gears are used. A three-gear train gives greater flexibility with the positioning of the camshaft in relation to the crankshaft. The pushrods act on rocker arms to operate the valves.

The example given here will use an overhead camshaft directly above the valves.

Picking Up the Drive

The final decision relating to camshafts is how the drive will be taken from the crankshaft. There are two alternatives – either from the front or the rear – and the choice will have considerable implications for the engine design.

With a single-cylinder engine, if the camshaft is driven from the front of the engine, a simple overhung crankshaft can be used. Among other things, this means that the big end of the connecting rod does not have to be split, making for easy crankshaft and connecting rod construction.

If the drive is taken from the rear of the crankshaft, it is considerably more complicated. Either the crankshaft has to be extended, with another web at the rear and a rear extension to the shaft providing the drive to the camshaft, or a separate 'follower' has to be provided. This would be driven off the main shaft crankpin and would need its own set of bearings. This layout means that the main crankshaft can still be an overhung type.

Toothed-belt camshaft drive on a single-cylinder engine, showing the belt, pulleys and tensioning arrangements. This type of drive is very simple and requires no lubrication. The layout for chain drives is very similar but needs lubrication and is much noisier.

In miniature engines, the most commonly used drive options are toothed belts or gears. Chains may be used, but toothed belts are easier to use and avoid the need for the complexities of lubrication. They are also somewhat more flexible regarding the distance between the crankshaft and camshaft, which may be of benefit. This is because the 'pitch' of the belt – the distance between adjacent teeth – is generally smaller, and belts are easily available down to 2mm pitch, while the smallest readily available chain-drive systems are 4mm pitch. Belts also take up a lot less room than the equivalent chain drive.

If the engine is a single-cylinder overhead-camshaft type, toothed belts are the easiest option because several gears will be needed to take the drive from the crankshaft to the camshaft and they have to be enclosed and lubricated. Toothed belts are readily available in small sizes, together with suitable pulleys, which are easily modified. As they run dry, any casing can be very simple and does not have to be oil-tight.

For the single-cylinder example given here, a toothed belt from the front of the crankshaft has been chosen, because this is by far the easiest method to design and make.

Crankshafts

Having made the decisions regarding the valve-gear drive, more of the crankshaft options can be fixed.

Ball-race bearings for crankshafts make for very easy lubrication.

Built-up crankshafts are favourable, but for smaller engines it is often as easy to turn them from the solid. This is one of the choices that may be influenced by the equipment available.

With a front-drive camshaft, the design can stick with a simple overhung crankshaft design with a toothed belt pulley between the bearings to drive the cams.

The crankshaft proportions can now be fixed, based on the cylinder bore size; this method seems to work and to give a reasonably well-balanced engine. Setting the main shaft diameter at around 0.4 times the bore seems to give a reasonable result with aluminium-alloy connecting rods and pistons. The crankpin diameter is normally set at about 0.3 times the bore but may be increased for higher-performance engines.

The crank web thickness should be approximately the same as the shaft diameter, with the web machined away to leave the traditional 'balanced' shape.

Plain big-end bearings can be bronze bushes or straight into a high-tensile aluminium connecting rod, both running on polished crankpins. Ball races tend to make big ends too bulky but needle-roller bearings may be used, pressed into the connecting rod – use the commercial inner races supplied for the bearings pressed on to silver-steel crankpins, and make the pin slightly larger to compensate.

The length of the shaft is, to a large extent, up to the designer and will be limited by what has to be fitted on to it (bearings, pulleys, propeller drivers/flywheels), and the use to which the engine will be put. For example, on a modern streamlined model aircraft, it may be advantageous to have the shaft longer to get the crankcase further back in the fuselage. If the engine is to be mounted across a narrow racing-car model, a short shaft is an advantage.

The basic crankcase may now be laid out, with the crankshaft in place, and clearance for the big end in the crankcase can be checked. This can be done without finalizing the position of the camshaft for pushrod or side-valve engines.

Method of Ignition
It may seem incongruous to consider the method of ignition so early in the process, but, because the compression ratio depends on which method is used, it is not possible to fix the compression ratio and hence the cylinder length until this has been decided.

There are three options: compression ignition (and I am aware of only two successful compression-ignition four-stroke engines), spark and glow plug. The spark and glow-plug methods differ in several respects, most obviously in the way they work and in the fuel used. The glow-plug system uses a hot platinum-based coil to ignite the methanol/fuel mixture, with the point of detonation dependent on the fuel, plug grade and compression ratio. The coil is heated electrically to start the engine but is then self sustaining.

The spark system uses a normal spark plug and petrol, with the point of ignition controlled by the contact breaker (or electronic ignition system). This system needs a power supply at all times, which may be a disadvantage for model aircraft use where weight is usually at a premium. It also requires provision for a contact breaker or electronic ignition sensor, normally driven from the camshaft.

Compression Ratio and Cylinder Length
Next, you need to fix the cylinder length and hence the position of the cylinder head. In order to do this, it is necessary to fix the compression ratio to be used in the engine. The

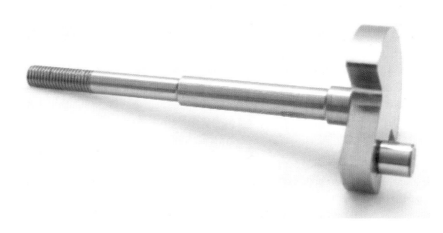

Single-cylinder overhung crankshaft using built-up construction.

compression ratio is dependent on the fuel and ignition method used.

Glow-plug engines running on methanol-based fuels need a compression ratio of around 10:1 (preferably a bit higher). Spark-ignition petrol engines can run with compression ratios as low as 5:1 or 6:1. For our type of medium-performance engine and running on spark ignition, I would set the ratio at 8 or 9:1. (I have built one engine with a compression ratio of 9:1, which will run on either glow-plug or spark ignition, but it does not tick over as reliably on glow plugs.)

Having set the compression ratio, it is necessary to work out the cylinder length to provide this ratio. 'Compression ratio' is defined as the ratio of the cylinder volume with the piston at bottom dead centre to that at top dead centre. For a flat cylinder head and piston it can be calculated closely enough using the following formula (this does not take into account the effect of the valves or spark-plug volume):

If the stroke is X
and
the distance between the piston crown and the cylinder head at top dead centre is Z
and
the bore is Y
then the compression ratio (CR) is

$$CR = \frac{X+Z}{Z/(CR-1)} \qquad CR = \frac{X+Z}{Z}$$

Since we need to calculate Z, this can be expressed as

$$Z = \frac{X}{(CR-1)}$$

This will allow you to set the height of the top of the cylinder and then move on to the cylinder head, the bottom of which is now fixed at this position.

Cylinder Head

If overhead valves are used, the cylinder head can be one of the most time-consuming parts to design. The basic problem is fitting

everything in. Having settled on the number of valves (two), this is the first place to start. Position the centres of the valves on a plan of the cylinder head on equal sides of the centre, ensuring that there is clearance between the valve heads and the cylinder walls, but leaving the maximum distance between the valve centres. This then fixes the position of the rocker-arm ends or camshaft and these are the next items to consider.

Valve Lift
The valves have to be lifted by the valve gear in order to allow fresh mixture in and exhaust gases out of the cylinder at the correct times. There is no point in having excessive valve lift; if the valve is lifted by an amount equal to one-quarter of the valve diameter, then the open area round the valve head is equal to the area of the valve port and therefore will pass the same amount of gas. Having established the valve diameter, you can define the lift required. Some builders allow for the fact that the valve stem occupies part of the port and increase the valve lift slightly; this does not seem to make much difference to any engines, other than very high-performance ones.

Once you know the valve lift required, the next thing to consider is cam design and valve timing.

Valve Timing
Valve timing (*see also* Chapter 1) has a significant effect on the performance and handling of an engine. In fact, engines can be changed dramatically by altering the valve timing. 'Valve timing' is the definition of the points at which the exhaust and inlet valves open and close, relative to the position of the crankshaft.

It is usual to express valve timing in relation to the top and bottom dead-centre positions of the crankshaft.

Cam Design

Once the valve timing and lift have been fixed it is possible to set out the cam layout.

The cams rotate at half engine speed, which

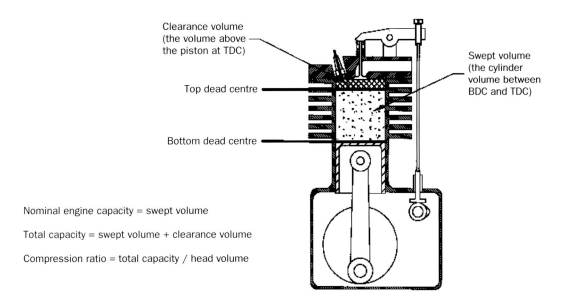

Clearance volume
(the volume above
the piston at TDC)

Swept volume
(the cylinder
volume between
BDC and TDC)

Top dead centre

Bottom dead centre

Nominal engine capacity = swept volume

Total capacity = swept volume + clearance volume

Compression ratio = total capacity / head volume

Compression ratio calculation on a single-cylinder engine.

means that the valve-open periods must be divided by two in order to set the valve opening and closing points on the cams. For the example engine being used here, the inlet valve is open from a point 20 degrees before top dead centre to a point 60 degrees after bottom dead centre; in other words, it has a valve-open period of 260 degrees. In order to achieve this, the cam must have an 'open period' of 130 degrees.

Knowing the lift required, you can now lay out the cam design. This is very easy with a CAD program but may require some trial and error with pencil and paper.

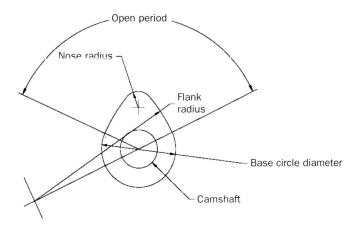

Open period

Nose radius

Flank
radius

Base circle diameter

Camshaft

Cam layout, showing the main dimensions used to define the profile.

The cam shape will depend on the type of tappet used. If a roller tappet is used, the cam flank can be flat – meaning that the cam can be cut easily using a rotary table in the mill. This type of follower is typically found on slow-speed, open-crank horizontal engines.

Use of the more normal flat tappet requires curved flanks on the cam in order to avoid shock loads on the valve gear.

Cams can be cut using several methods (*see* Chapter 12).

Exhaust and Inlet Cam Relationship

Once the cams have been designed and laid out, it is necessary to determine their correct angular relationship to each other on the camshaft. Again, this is easier to follow using the type of timing diagram with two strokes in terms of crankshaft angle (*see* panel, right).

For the example engine, the angle between the cam noses is 105 degrees. This is calculated by taking the total of the valve open periods, dividing by two (to get the mid point) and subtracting the total valve overlap and dividing the result by two to get the cam angle.

Remember that, at the start of the overlap period, the exhaust cam is leading in the direction of rotation.

Lubrication

All the bearings and sliding faces in an engine need to be provided with adequate lubrication in operation. The method by which this is provided will depend on the type of engine being considered.

For two-stroke engines the oil is mixed with the fuel and, as the mixture travels through the crankcase, oil is deposited on the bearings and other parts of the engine.

For a slow-revving open-crank horizontal engine, the lubrication is usually by means of oil pots linked to the bearings, which allow oil to reach the bearings by gravity or wick feed, often regulated by some form of flow-control valve. This method can also be applied to the

Four-Stroke Engine Timing – Design Considerations

The example given here relates to a medium-performance engine; this implies timing that provides flexible running and easy starting.

Inlet Timing

The inlet valve controls the admission of fresh fuel/air mixture into the engine during the inlet stroke of the cycle. The very basic requirement is that the valve should open at top dead centre on the inlet stroke and close at bottom dead centre. This would get mixture into the engine but not very efficiently; the events in an engine take place many times per second and at these speeds the fuel/air mixture has a certain amount of inertia, which affects the valve timing.

It takes time to accelerate the column of gas in the inlet tract and, while this is happening, the piston is moving down but no mixture is entering the cylinder. At the bottom of the stroke, the column of gas has maximum inertia; if the inlet valve is left open after bottom dead centre, this inertia will force more mixture into the cylinder, increasing the efficiency of the engine. It is normal therefore to open the inlet some time before top dead centre and to close it after bottom dead centre. The amount depends on the performance required and affects the ease of starting and flexibility of the engine.

A medium-performance engine will have the inlet opening around 15 to 25 degrees before top dead centre and closing between 50 and 60 degrees after bottom dead centre. These figures are a typical example; the values will vary from engine to engine.

Exhaust Timing

The exhaust timing is affected by inertia in the same way as the inlet timing is. In the case of the exhaust, it needs to open before bottom dead centre on the power stroke so that the exhaust gases can accelerate in order to travel through the exhaust tract. At top dead centre the inertia is used to force more gas out of the exhaust port and at this point will be assisted by the fresh mixture entering the cylinder. This gives good scavenging, meaning

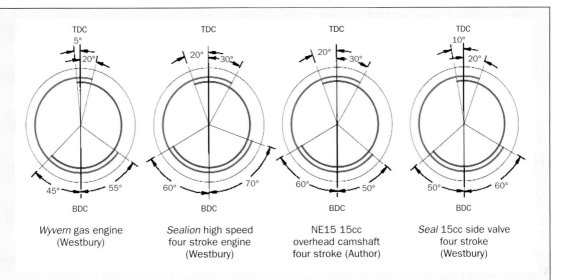

Valve timing diagrams for several different engines.

Wyvern gas engine
(Westbury)

Sealion high speed
four stroke engine
(Westbury)

NE15 15cc
overhead camshaft
four stroke (Author)

Seal 15cc side valve
four stroke
(Westbury)

that the burnt gases are fully expelled from the cylinder, allowing the maximum charge of fresh mixture.

For a medium-performance engine, the exhaust would be expected to open between 45 and 55 degrees before bottom dead centre and close 25 to 40 degrees after top dead centre. The exhaust opening period may be made a little longer than the inlet for slightly improved performance, but this is not essential.

The period round top dead centre when both valves are open is called the 'overlap period' and will be substantial on very high-revving high-performance engines. Such timing also makes the engine difficult to start and somewhat inflexible in operation.

Timing diagrams show the valve timing for a range of engines by different designers.

Timing Diagrams

The valve timing of an engine is expressed using a timing diagram, which shows the total cycle and valve opening and closing periods. There are two versions of timing diagrams,

the most common showing the four strokes overlapping. Unfortunately this type is not the easiest to understand because it does not show the power and compression strokes as separate entities; it only shows one bottom dead centre and one top dead centre.

The second type of timing diagram shows the four strokes as separate entities, with two pairs of TDC and BDC positions at 45-degree intervals, giving a quarter of the circle for each stroke.

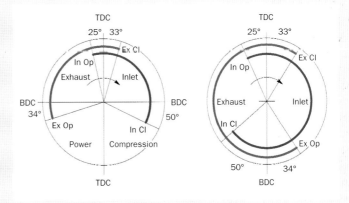

Different types of timing diagrams. Right, by crankshaft angle, left by camshaft angle.

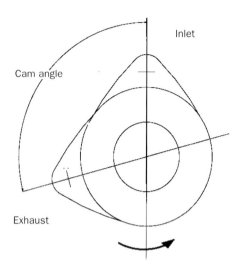

Inlet and exhaust cam relationship. The exhaust cam always leads in the direction of rotation and the angle is calculated from the timing diagram.

cylinders of such engines. Camshaft and other smaller bearings are commonly fitted with oiling points and are oiled manually as needed. All these are total loss systems.

For more sophisticated four-stroke engines there are several options available. These include gear or plunger pumps driven off the crankshaft or camshaft, splash lubrication, crankcase pressure feed, and mixing oil with the fuel, as in two-stroke engines.

Oil Pumps

Use of an oil pump obviously adds complexity to an engine but can provide oil feeds to all the bearings and, because the oil is circulating, this system can also provide extra cooling for the engine.

The options available are small gear pumps or plunger-type pumps. Plunger-type pumps are more difficult to fit into the engine because they are generally more bulky and the reciprocating drive is trickier to arrange. They are often driven by a skew gear-driven shaft from the camshaft and located at the bottom of the sump. They may also be cam-driven,

sometimes with a spring to provide the return stroke.

A gear pump is the type used on a full-size engine and can be driven off the end of the camshaft (particularly for side-valve engines) or be a separate unit driven from the timing belt or camshaft drive train.

For spark-ignition engines with a vertical distributor drive shaft, the oil pump can be driven off the lower end of this shaft and located at the bottom of the sump. The benefit of this is that the pump is submerged in the oil.

Gear pumps work well, but need to be accurately made. They will usually provide more oil than is required so a pressure relief valve needs to be incorporated into the delivery pipe.

Splash Lubrication

This form of lubrication has been used for many engines and for a medium-performance engine will provide adequate lubrication for the bottom end of the engine and cylinder. The principle is that the sump is filled with oil to a level at which it covers the bottom of the connecting-rod big end. As the crankshaft rotates, the big ends dip into the oil and throw it round inside the crankcase. The big ends

Gear pump showing the internal arrangement. This is a water pump, but the construction for oil pumps is the same.

often have a small 'dipper' with a hole connecting to the bearing to force oil into the bearing. For side-valve engines, the camshaft will also be lubricated and if the camshaft drive gear casing is connected to the crankcase, the drive will be lubricated by the oil mist. This method of lubrication is used in the Edgar Westbury *Seal* four-cylinder design, among others. Sometimes, a curved tray is fixed under the big ends with a central hole under each big end, which allows the oil to feed up from underneath, to be picked up by the big end.

One problem with splash lubrication occurs when considering lubrication for the valve gear on overhead-camshaft engines. One way round this problem is to take a breather from the crankcase up to the cam box with the vent from the engine in the top of the cam box. Sufficient oil mist will normally be pushed up the breather to provide lubrication of the cams and tappets.

Crankcase Pressure Feed

This method of lubrication uses the pressure variations in the crankcase to suck oil in from an external tank, often via the crankshaft main bearing and big end. A light one-way valve is provided in the crankcase, which, as the piston goes down, allows air to be expelled from the crankcase; when the piston rises, the valve closes. The rising piston creates a pressure drop in the crankcase, which can be used to suck oil in from the tank.

In order for this to work effectively, the crankcase must be sealed, with the one-way valve being the only vent.

For an overhead-camshaft engine the one-way valve should be in the cam box with a breather connecting crankcase and cam box, so that oil mist is pulled into the cam box from the crankcase.

The oil normally enters the crankcase via oil-ways in the crankshaft leading through to the big-end bearing from which the oil is flung round the inside of the crankcase. A small hole is provided in the crankshaft which connects with a corresponding hole in the main bearing as the crankshaft rotates. This is timed to be at the point when the pressure in the crankcase is at its lowest, so that oil is sucked through into the engine.

If the engine uses ball-race crankshaft bearings, a short section of plain bearing must be provided for the oil connection.

This method of lubrication is most commonly used on single-cylinder engines because of the greater pressure variation in the crankcase of such engines. It has also been used in the past with multi-cylinder in-line engines, which do generate (smaller) pressure variations because of the geometry of the relationships between the events in each of the cylinders.

Oil/Fuel Mixture

This method of lubrication is similar to that used in two-stroke engines. Four-stroke engines can have two variations.

In the first, the fuel mixture, with oil, is taken into the cylinders via the crankcase, thus lubricating the engine internals. Oil mist is used to lubricate valve gear and the crankcase must be sealed as in two-strokes.

In the second variation, the fuel mixture is taken directly into the cylinder and the bottom-end lubrication relies on the small amount of oil that goes past the piston when the engine is operating. A small amount of oil is often inserted into the crankcase before the engine is started to provide the initial lubrication.

This method is probably best with ball-race or needle-roller bearings, but is also used with plain bearing engines.

Once the lubrication method has been chosen, it is possible to lay out any pumps and their positions together with the basic drive.

Carburettors

Carburettors on small engines have been made in many types but these days the most common is the barrel-type carburettor, which is reasonably easy to make and to set up. It consists of a cylindrical barrel, which can rotate inside the carburettor body. The air passage is drilled

Simple barrel-type carburettor for a four-stroke engine.

through the body and barrel at right-angles to the rotation axis of the barrel and on the centre line. As the barrel is rotated, the air passage is throttled to a greater or lesser extent.

The fuel is mixed with the air by means of a small jet in a spray bar that is located across the barrel parallel to the axis of rotation, with the jet facing into the engine. The air flow past the jet sucks fuel out; the fuel then mixes with the air and travels into the engine. The amount of fuel is controlled using an adjustable needle valve allowing the fuel/air mixture to be adjusted to suit operating conditions.

The minimum throttle setting is generally set using an adjustable throttle stop screw that prevents the barrel from rotating past the minimum throttle opening set.

More complex variations of this type of carburettor include the addition of slow-running needle valves and also air-bleed screws to fine-tune the mixture at low throttle settings.

In terms of the design parameters to be considered, the most important is the size of the air passage (the choke) to be used. If the passage is too small, the upper speed range of

the engine will be restricted, although it will make it easy to start and less critical on the settings. A large choke size will make the engine more critical to set up and less easy to start, and will also affect the slow-running ability of the engine.

As a guideline, for a low-performance slow-revving engine, a choke size of approximately one-eighth of the cylinder bore size is sensible. For the medium-performance example given here, one-fifth is sensible, and for out and out high-performance engines, sizes up to one-third of the bore size are used.

One thing to note about this type of carburettor is that it is not a 'compensating' type – it does not automatically alter the mixture to suit the load on the engine. Many attempts have been made to make compensating carburettors for miniature engines, with varying degrees of success.

A full-size car engine carburettor detects changes in the inlet vacuum and uses these to open or close the fuel needle valve, providing a richer (more fuel) mixture when the load increases, and going leaner when the load reduces. The problem with model sizes is that air-flow properties do not scale but those who wish to experiment on larger engines may get some success. The best-known carburettors of this type were made by Weber, Stromberg and SU, but with the advent of computer-controlled fuel injection, carburettors have all but disappeared from modern full-size engines.

It is not normal to use a float chamber with a barrel-type carburettor because it will suck fuel up from quite a large head and the mixture will remain more or less constant in normal use.

Some high-performance engines use a pressurized fuel system to ensure a good fuel feed in all situations. This is particularly common for model aircraft use.

The spray bar should be the minimum size possible and the jet size should range from 0.3mm for smaller engines of around 5cc capacity to around 0.7mm for 15cc engines. In any case, it is easy to experiment with jet sizes and the adjustable needle valve provides a wide range of adjustment.

The carburettor can then be set out. As with the cylinder head, the problem is often fitting everything in without making the carburettor look oversized.

Ignition Systems

There are three main types of ignition system used on miniature engines: spark, glow-plug and compression ignition.

Compression Ignition

Compression ignition is seen in the so-called diesel two-stroke engines that run on an ether/paraffin fuel mixture. The compression ratio is adjustable by means of a screw-adjusted contra-piston at the top of the cylinder. Some early engine designs had no compression adjustment and had to be provided with a very rich mixture for starting. Ignition takes place because the high compression ratio in such engines raises the cylinder temperature to the extent that the highly volatile ether ignites, causing ignition of the rest of the mixture.

The point of ignition is purely dependent on the compression ratio and fuel mixture.

Spark Ignition

Spark-ignition engines run on ordinary petrol, with the point of ignition controlled by the timing of the spark. The spark is normally generated by means of a coil that amplifies a low voltage from the battery into the high voltage needed for the spark to jump across the spark-plug electrodes.

Older engines used a set of cam-driven contact points acting directly on the coil low-voltage circuit to control the ignition. Nowadays, many builders use electronic-ignition systems, either with a contact breaker or an electronic switching circuit driven by a magnet on the camshaft acting on a Hall Effect transistor. These electronic systems offer many advantages to the engine builder –

they are much more reliable than a small contact breaker and can be fitted into a smaller space.

In both systems, the contact breaker is driven directly from the camshaft for four-stroke engines, while two-stroke engines have the contact breaker as part of the crankshaft. However, for a single-cylinder four-stroke engine the ignition is sometimes driven from the crankshaft, which gives two sparks per cycle. Since the redundant spark occurs at the end of the exhaust stroke, it does not cause any problems.

Small ready-made systems or kits of parts are available from many suppliers and those with electronics expertise can design their own.

One important point to note with any spark-ignition system is that the spark occurs when the electrical supply to the coil is broken, not when the points close. The sequence is as follows: the points close, the coil is charged up, the points then open, generating the spark and causing ignition. The ignition cam must therefore close the points for long enough to charge up the ignition

Contact breaker assembly from a single-cylinder two-stroke engine. The cam is ground directly into the crankshaft and the moving contact is the insulated one in this example.

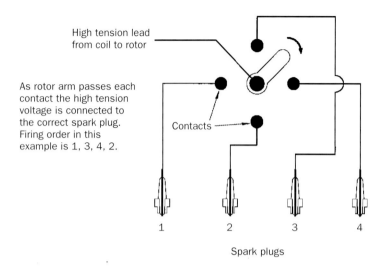

High tension lead from coil to rotor

As rotor arm passes each contact the high tension voltage is connected to the correct spark plug. Firing order in this example is 1, 3, 4, 2.

Contacts

1   2   3   4

Spark plugs

The operation and connections of a distributor for a four-cylinder engine.

circuit. A closed period of 60 to 70 degrees is reasonable and ensures that the coil will not overheat and that the battery consumption is kept to the minimum.

With Hall Effect systems, the time the magnet is activating the sensor controls the closed period. This depends on the diameter of the magnet and its rotating radius. It is possible to use a radius down to 10mm with a magnet of 2.5mm diameter with no problems.

With spark ignition some means of providing control over the timing advance and retard must be provided. This is normally by means of rotating the contact or sensor assembly around the camshaft, thus altering the point at which the spark is triggered in relation to the engine timing. Some builders also provide some form of automatic advance and retard linked to the throttle control. This is because the ignition needs to fire earlier when the engine is running fast because of the delay in actual ignition of the mixture. This is also done electronically in some sophisticated electronic-ignition systems.

With multi-cylinder engines, a means of getting a spark to the correct cylinder in turn is needed. This was always by means of a distributor in the past but some builders now use separate coils and electronic systems for each cylinder, which is in line with modern full-size practice but can be expensive if you buy the components. The distributor consists of a rotating contact that connects the high-tension lead from the coil to a contact connected to the relevant spark plug. It is normally incorporated into the contact breaker and driven from the camshaft.

One point to make about using Hall Effect sensors for multi-cylinder engines is to avoid locating the sensor inside the distributor because the high voltages and sparking that take place inside such small distributors may well destroy the sensor.

Glow-Plug Ignition
Glow-plug ignition is similar in some respects to compression ignition in that the ignition is controlled partly by the compression of the mixture in the cylinder. The difference is that the actual ignition is caused by a hot platinum-based coil in the cylinder – the glow plug.

The coil is heated up by a low voltage supply (1.5–2 volts) for starting, although, once the engine is running, the action of the fuel and combustion keep the coil hot, allowing the supply to be disconnected. The fuel used is a methanol-based fuel with a small amount of nitro methane added to improve running and starting.

This type of ignition obviously does not require any contact breaker or coils, which means that engine construction is simplified. It can however be more difficult to get an engine running well because the ignition point depends on three things: the compression ratio, the fuel mix used and the heat rating of the glow plug.

WORKSHOP CONSIDERATIONS

One of the factors that must be taken into account when making design decisions is the availability of equipment. If consideration is taken of this during the early design stages, problems can be avoided later in the construction process.

As an example, a 15cc single-cylinder pushrod engine may be built entirely on a small Hobbymat lathe and milling machine. For easy machining, the cylinder jacket is in two parts, to reduce the overhang in the lathe and the crankcase is split on the vertical axis so that it may be mounted in the four-jaw chuck to bore the bearing housings. If such design considerations are not taken into account at an early stage, other external facilities will need to be located in order to machine all the parts.

DRAWINGS

Once you have made all the basic design decisions the time will come when some sort of working drawing will need to be produced. Obviously, the number and complexity of the drawings will depend on the engine being built, but also on whether the engine is a one-off for the builder, or is intended to be passed on to others or entered into a competition.

In the past the only option was to use pencil and paper and this often resulted in two cycles of drawings. The first set comprised the design drawings, in which all the details were finalized and the layouts established. These were followed by a set of working drawings, which were often not finalized until the prototype had been built.

Computer-Aided Design

These days, model-engine builders have the use of computer drawing systems, which make the task of producing good-quality, accurate drawings much easier. Computer-aided design (CAD) systems are easily available and basic systems can be found free on the Internet. Even a fully functional system including three-dimensional (3D) drawing facilities can be found for well under £100 at the time of writing.

It is important to find a package that will do the job, and then to spend time learning it thoroughly. A good way to learn is actually to design and draw an engine. It will be slow at first, but at the end you will have a good knowledge of your chosen package.

The big advantage of CAD systems is that as the design progresses, any changes to parts already drawn can be made very easily. If you are not sure which of two particular choices is best you can save a copy of the first option and then draw a second option before deciding which to use.

The other advantage of these systems is the ability to copy parts. For example, if you are drawing a four-cylinder engine, once one cylinder is drawn it can be copied three times and all four cylinders are done.

Some Basic Rules

Following a number of basic rules will save time and effort:

1. set some standards for your drawings. This is essential if they are to be published;
2. decide on the projection system used and keep to it. If you do not understand the difference between first- and third-angle projection, a search on the Internet or in the local library will provide the information;
3. most packages use a system of 'layers'. Using

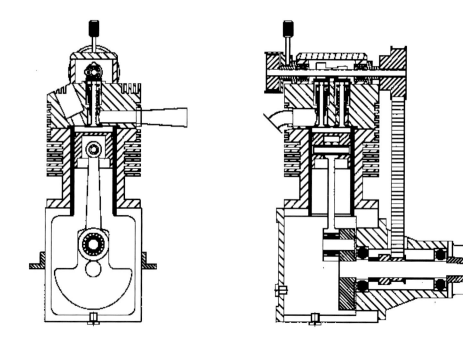

ABOVE: *CAD-produced cross-sectioned general arrangement of a 15cc four-stroke engine.*

LEFT: *Another view of the author's 15cc OHC engine. All the major parts in this design are machined from the solid.*

them can make life a lot easier. For example, you can use separate layers for the main drawing lines, dimensions, construction lines, hatching, hidden lines, centre lines and PCD lines. Each of these can be included or excluded from the drawing view with a click or two of the mouse. In addition, in most packages, the colour and type (solid, broken, centre) of any line can be set for the relevant layer;

4. do not draw everything by 'dragging' with the mouse and positioning by eye. Make use of the 'snap' facilities to join things and use typed-in values to set line length, angle, diameter, and so on. This method is much more accurate;

5. draw the general arrangement (GA) first. The advantage of this is that parts can be fitted to each other so you avoid a situation in which parts do not match up. Also, as you build up the GA, you can copy individual parts for production of the detail drawings later. For example, once a basic piston side view is in place it can be copied and saved for later production of the detail drawings.

Most modern computer systems are more than adequate to run a CAD program but if you want to produce many drawings you could invest in an A3-size printer. They are not that expensive these days and the larger sheet size provides much more flexibility in your drawing. It is amazing how much room even a small engine takes up when drawn full size in two views.

WHAT TO BUILD FIRST?

To sum up, the answer to this question is up to you but it might be a good idea to start with a single-cylinder engine, and preferably a proven design such as one by Edgar Westbury. In this way, you know that the engine will run and also that there will be help around from previous builders.

A popular design by David Parker, the V-twin Vega four-stroke engine is slightly more complex than a single, but well within the capabilities of a competent model engineer.

The choice between a two- or four-stroke engine is really a personal one. A two-stroke engine is simpler but cutting the cylinder ports can be tricky. The four-stroke has the complexity of cams and camshafts but these need not be a big problem – cams can be cut on the milling machine or lathe (using a suitable easily made jig) and set on the camshaft using Loctite.

Perhaps the best option therefore for a first engine is a single-cylinder four-stroke of between 10 and 15cc with pushrod valves. Choose a size suitable for your equipment. If the machining can be handled comfortably in your own workshop you only have to worry about the engine, not how you are going to fit it in the lathe.

3 Workshop Equipment and Tooling

INTRODUCTION

Any experienced model engineer will almost certainly have all the tools needed to build internal combustion engines in his or her workshop. This chapter will cover the basics, as well as some additional equipment that will make life easier.

MACHINING EQUIPMENT

Lathe

A lathe is obviously essential and any model engineer's type of lathe is suitable. Size will depend on the size and type of engine being considered.

There is one important point to make for those interested in building models of open-crank 'gas' engines: these have large flywheels and therefore the turning facilities must be able to handle a larger diameter than with other engines.

It is possible to machine, for example, a 15cc capacity single-cylinder engine on a small 60mm swing centre lathe but this will involve some design considerations to ensure that it may be done. Any lathe from 3½-inch swing upwards should be suitable for most engines below 15cc capacity.

If a milling machine is not available then a vertical slide is needed in order to carry out many of the necessary milling operations. In this case, the lathe probably needs to be larger to handle this type of work.

A collet chuck is also useful for holding smaller items such as valves.

Milling Machine

A vertical milling machine of some type will make the construction of I/C engines much easier, particularly when making multi-cylinder engines. Any type of small mill will be suitable; the V-column type may be preferable because the head can be moved up and down without losing the horizontal positioning.

A rotary table (or dividing head) and boring head are useful accessories to use with the mill and the fitting of digital read-outs to all axes will make complex machining very easy. If cams are to be cut in the mill, then a digital read-out should be fitted to the vertical axis for this operation.

An angle plate or vice will aid setting up to drill angled holes such as the spark-plug holes in overhead-valve cylinder heads.

A table size of around 175×600mm is a sensible lower limit but do get a mill with the maximum possible height under the quill. This is particularly important when using the rotary table.

If a milling machine is not an option, then a good-quality pillar drill is essential.

ABOVE: A lathe set up for milling with vertical slide mounted on saddle.

RIGHT: A compact milling machine with rotary table and chuck, and equipped with digital read-outs on all three axes.

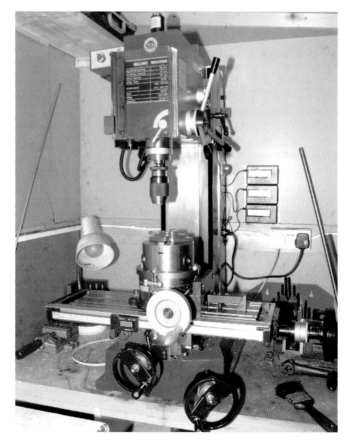

MEASURING EQUIPMENT

In addition to the usual rules, set squares and marking-out equipment, micrometers and vernier gauges of suitable size are essential. These days, digital versions of these are avail-able very cheaply and have the advantage of being able to zero at any position; this is very useful when machining parts to fit others already made. An internal micrometer for the sizes of cylinders being considered would also be of benefit.

A height gauge and surface plate are useful for marking out and also for setting up parts for drilling holes from other parts, or setting up parts in jigs.

A good substitute for a surface plate is a small sheet of plate glass.

One item that may not be available is a good-quality vernier protractor, which makes setting up some of the odd angles found in I/C engines much easier.

JIGS AND FIXTURES

During the construction of even a simple I/C engine it is likely that some basic jigs will be needed. For example, a drilling jig machined at the same setting as the cylinder bolt down holes will be used to drill the corresponding holes in the crankcase.

In the case of more complex multi-cylinder engines, jigs are used to ensure repeatability when machining identical items. This is particularly true of radial engines where jigs are commonly used for the cylinder heads, cylinder bolt holes, valve housings and, often, many other items.

Several jigs may be needed for each part and the builder will have to decide how far to go down this path. Where several identical items with angled drilling or milling operations are needed, jigs will always save time in the end. Similarly, if you are building a batch of several identical engines, suitable jigs and fixtures will be essential.

The use and type of jigs should be considered at the design stage because attention to this point early on can make the design and construction of both the jig and its associated component(s) easier.

If digital read-outs are fitted to the milling machine, the need for some jigs will be reduced. For those using CNC facilities, fixtures are used to ensure accurate positioning of items for machining.

DIGITAL READ-OUTS AND CNC

Many constructors have digital read-outs fitted to their machines or have computer numerical control (CNC) equipment. Both of these make the machining of complex shapes easier (or even, in the case of CNC, possible).

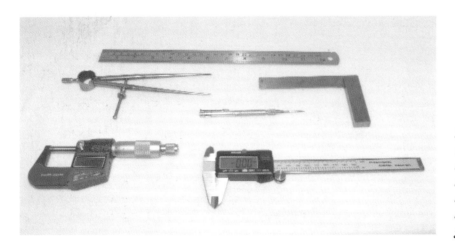

Some essential basic measuring equipment: rule, dividers, scriber, digital micrometer and digital vernier gauge.

The commercial digital readout system fitted to the author's milling machine. The bar is upside down so that the cable exits underneath and the readout is protected by an aluminium angle shield.

Digital Read-Outs

The fitting of digital read-outs to machines, particularly milling machines, makes the setting up and machining of parts for miniature engines much easier. Such equipment will often remove the need for jigs and is a great aid to repeatability – an essential for multi-cylinder engines, because the settings for the first of a set of identical items can be noted down and then used for the rest of the set.

Add-on read-outs can be obtained from many suppliers and it is certainly worth considering fitting one to the milling machine.

Some computer programs calculate the milling offsets for cam profiles and those wishing to make use of these will need to fit a digital read-out to the vertical axis of the machine at least.

Computer Numerical Control

Small CNC machines are now available from many suppliers and can be beneficial to those making large complex engines from the solid. Using such facilities will mean learning some new skills – such as a form of computer programming – not normally associated with model engineering, but they will make the machining of parts for multi-cylinder engines easier. Certainly, consistency will be assured.

Those interested in finding out more about the subject will find information on the Internet, along with several suppliers selling parts for the conversion of existing machines. In addition, for those producing CAD drawings for their engines, data from the drawing program can be used as input to the machining process.

Add-on read-outs can be obtained from many suppliers and it is certainly worth considering fitting one to the milling machine. Such readouts will need mounting brackets made to fit the particular machine and can be used with remote readouts if required.

Part II: Four-Stroke Engines

4 Crankshafts, Bearings and Flywheels

CRANKSHAFTS

The crankshaft is the heart of an engine and is also one of the most highly stressed components, requiring care in its manufacture.

Materials
Contrary to widely held belief, it is not necessary to use exotic materials for the crank-shafts of miniature engines, unless extra high performance with absolute minimum weight is sought.

Single-Cylinder Engines
For single-cylinder engines, a mild steel such as

EN8 is suitable. EN1A free-cutting steel may also be used without any problems. For slow-revving single-cylinder horizontal engines, EN1A can be used throughout, with the crankshaft and big-end bearings of bronze. In this case, the camshaft is often driven via a pair of skew gears from one side of the crankshaft. These gears do not need to be hardened and are generally located with a pin in the crankshaft.

For higher-revving engines, the tougher EN8 grade is best if plain bearings are to be used. Free-cutting mild steel should be avoided if the crankshaft is of silver-soldered built-up construction.

Single-cylinder horizontal gas engine crankshaft.

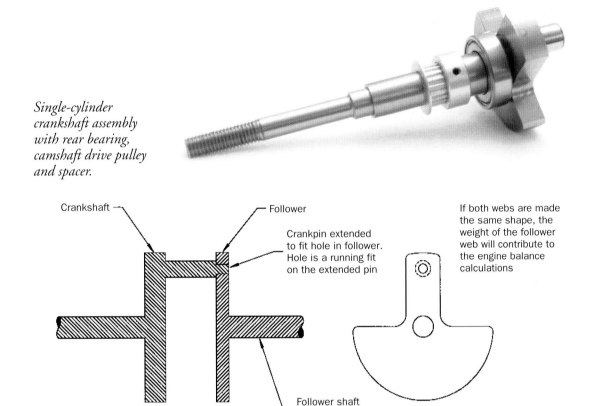

Single-cylinder crankshaft assembly with rear bearing, camshaft drive pulley and spacer.

Single-cylinder crankshaft with follower crank to drive valve gear. An extended crank pin engages a hole in the follower when assembled.

One possible combination is a composite construction with an EN8 main shaft, an EN1A web and an unhardened silver steel crankpin.

If the camshaft drive is from the rear, then a follower can be used to drive this. This is made in the same way as the crankshaft proper. It runs in its own bearings in the crankcase rear and is driven from an extended crankpin. The other option is to make the crankshaft and follower as one unit and to use a split big-end bearing.

If the camshaft is driven from the front of the engine, no follower is needed and a simple overhung crankshaft is required. In this case, a silver-steel crankpin can be used because, with a simple overhung crank, the pin is only supported at one end.

Multi-Cylinder Engines

With multi-cylinder engines, the crankshaft is often milled and turned from the solid but it can be built up if required. The same materials are used as for single-cylinder engines but multi-cylinder engines will typically have split big ends, although there are engines that have roller-bearing crankshafts that are pressed together in jigs to ensure accurate alignment.

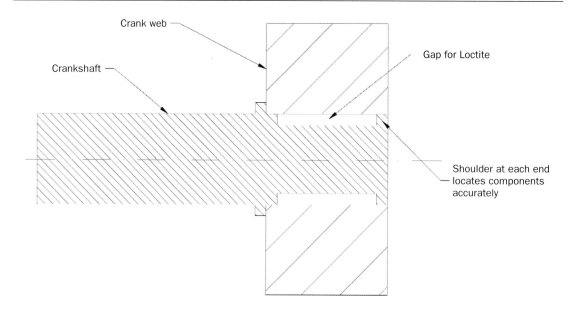

Crank web

Crankshaft

Gap for Loctite

Shoulder at each end
locates components
accurately

Using flanges for accurate location of Loctite joints. The gap must be in line with the adhesive manufacturer's recommendations and must be filled with adhesive for a strong joint.

Crankshaft Construction Techniques

The first decision to be made is whether to build up the crankshaft or to machine it from one piece. This decision is often influenced by the machinery available because, for example, it may not be possible to set a crankshaft over far enough to turn the crankpin in a small lathe.

Machining

Multi-cylinder engine crankshafts are typically machined from the solid with the shaft set between centres. In this case, a suitable length of bar is marked out and the waste metal between the webs is sawn and milled out first. For a typical in-line engine, the bar can be a suitable size of rectangular section that will reduce the amount of metal to be removed.

If using bright mild steel, the bar should be annealed before use to avoid distortion during the machining process. Black bar can be used 'as is'.

The bar is then set up between accurately marked and drilled centres for turning the big-end bearings and the main shaft. When turning the big-end bearings it is advisable to insert supporting spacers between the crank webs to assist in taking the pressure of the centres; this process will protect the shaft from bending. The potential problem with this method is that the lathe tool has to protrude a long way from the tool post, which means that very light cuts must be taken.

Some builders build crankshaft-grinding jigs for finishing the bearings of such shafts but with careful machining this is not essential. Many of the multi-cylinder designs of Edgar Westbury had crankshafts milled and turned from mild steel with the big-end bearings running directly in the aluminium-alloy connecting rods.

Assembling Built-Up Crankshafts

Built-up crankshafts become a series of simple turning jobs, with the separate big ends, webs and shaft sections being assembled after they have been machined. In this case, jigs may be used to ensure that all the parts are machined accurately.

There are three options for assembling built-up crankshafts: silver-soldering, press-fits and the use of modern adhesives such as Loctite. Each has its supporters and all are used widely.

Silver-Soldered Crankshafts Before the advent of adhesives such as Loctite, silver-soldering was the only alternative to the use of press-fits. If silver-soldering is to be used, then free-cutting steel should be avoided; it contains a proportion of lead, which can prevent the solder taking to the metal surfaces and may result in an unsound joint.

If using silver solder, the components should not be finish machined but should be left slightly oversize for cleaning up after the soldering is complete. This is because the heat of the soldering may cause (some) distortion.

This also highlights the point that, since the crankshaft will have to be finished after soldering, it may be necessary to leave centres on the parts in order to facilitate this. In this case, builders may prefer to turn the shaft from the solid from the start.

A small gap must be left in the joints to allow the silver solder to flow through the joint and it goes without saying that the surfaces must be thoroughly cleaned before soldering.

Press-Fitting Crankshafts Press-fit assembly of crankshafts avoids the problem of distortion but does require accurate turning to achieve the correct fit. The shaft must be 0.00075in per inch (0.02mm per 25mm) of diameter plus 0.0005in (0.125mm) larger than the hole diameter to give the correct fit.

It is also common to drill through the joints and to press in securing pins once the main items are together.

When pressing the shaft components, if possible, use a proper press to ensure correct alignment. If this cannot be done, it is worth making up simple jigs to help ensure that everything is lined up correctly.

Bonded Crankshafts These days, the use of modern industrial adhesives such as Loctite is very common throughout industry and model engineers are taking advantage of this in their work. Such adhesives aid easy assembly and prevent possible distortion of the crankshaft.

Two things are critical to the success of a joint made this way: the correct fit between the mating parts (to allow room for the adhesive) and absolute cleanliness. The fit varies between adhesives but reference to the relevant data sheets will provide the information needed. Cleanliness is taken care of by careful degreasing, followed by use of the proprietary cleaner for the adhesive in question. It is important that the mating surfaces should not be touched after cleaning. It is also sensible to insert pins for additional security once the adhesive has cured.

Because the fit between mating parts is not a tight one, it is essential to use jigs or to design the parts with a locating flange to ensure correct alignment is maintained whilst the adhesive cures. One easy way of ensuring that the web and shaft are in alignment is to set up the web in the lathe and then use the tailstock chuck to hold the shaft.

BEARINGS

The bearings used in model engines are of two main types, plain and ball or roller bearings. The type of bearings chosen will have important implications for the lubrication method selected. In fact, the choice of bearings may depend on the method of lubrication, and vice versa.

Plain Bearings

Plain bearings are commonly used in all areas of miniature I/C engines; indeed, in the early days of model engine development, they were the only option. They are still used for parts such as the gudgeon pin and often for big-end bearings, particularly with split big ends in multi-cylinder engines. Such bearings are typically made from a bronze alloy, although gudgeon pins and big ends often run directly in aluminium-alloy connecting rods.

Another option is to use sintered (Oilite) bearing inserts, which are pressed into their housings. If these are used, the bearings should not be reamed to size after installation because this will close the oil pores in the bearing material, leading to premature failure. For the same reason, Loctite should not be used to secure such bearings. In any case, they are designed to be a press-fit in the nominated reamed hole, which causes the bearing to contract to the correct fit for the designed shaft size. Neither is it advisable to use such bearings for high intermittent load applications such as big ends and main bearings. They are ideal for camshafts and similar situations.

Where plain bearings are used, it is essential to ensure that the shaft has a very good finish, to avoid premature wear, and that adequate lubrication is provided.

When using plain bearings as the main bearing for an overhung crank (as in two-stroke engines), the bearing bush should be pressed into the crankcase and then bored to size after fitting.

Plain main bearing from a small engine. The bearing bush is pressed in and should be reamed or bored to final size after fitting.

With long bearings, a spiral oil groove may be beneficial to good lubrication.

If using plain bearings running directly in the connecting rod, then the rod should be made of high-tensile aluminium alloy such as HE15, which is a very good bearing material when used with steel shafts. In this situation, it is possible to use unhardened silver steel for the crankpin.

Ball and Roller Bearings

Ball and roller bearings are widely available in very small sizes and are a very good option for miniature engines. They will operate happily with very little lubrication and, as long as there is some oil mist round the bearing, no problems will occur. This means that, if roller bearings and ball races are used for the big-end and main bearings, splash lubrication will be more than adequate. In fact, many engines have been made that use oil in the fuel and rely on some of that oil blowing past the piston to provide lubrication for the bottom end of the engine.

Bearings can be obtained as open (with no seals), shielded (with metal shields covering the balls), and sealed (with rubber seals completely sealing the balls).

One disadvantage with ball and roller bearings is that unsealed and shielded bearings are susceptible to the ingress of grit or other small particles. This means that care needs to be taken when fitting them and also when adding oil to an engine. If any bearing has a gritty feel when rotated, it should be cleaned thoroughly before fitting by washing in thinners to remove any particles.

This problem can also occur (although to a lesser degree) with shielded ball races, however, if grit gets in one of these it can be impossible to remove and the bearing will have to be scrapped. Sealed ball races do not suffer from this problem. They are sealed and lubricated for their working life, making them a good choice for main bearings.

Ball races need to be fitted correctly, with a push-fit of the inner on to the shaft and a light

An unusual built-up crankshaft for a four-cylinder engine, featuring large-diameter ball-race main bearings and needle-roller big ends. The main bearings are located in machined housings in a split crankcase.

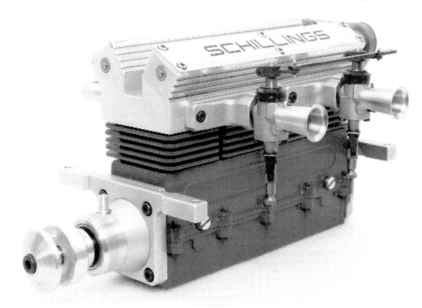

A view of the 40cc four cylinder twin overhead camshaft glow engine by Schillings. Variants of this engine were produced for several years and have a pressed and bolted built up crankshaft with needle roller big-end bearings.

press-fit into the housing. When pressing in bearings, use a press or the drilling machine to ensure the bearing is inserted in correct alignment. When installed, the shaft should turn freely with no sign of friction. If it feels tight, the bearing housing is too small and should be enlarged.

Needle-roller bearings also need to be fitted correctly. This is particularly true of the drawn cup type of bearing, which is designed to be pressed into a reamed hole that compresses the outer housing to the correct size.

Needle-roller bearings are also available with inner races, which can be fitted to a separate crankpin or crankshaft using a light press-fit or Loctite. If these are not used, then the shaft must have a very good finish and will need to be made of a tougher grade of steel.

One other disadvantage with ball and roller races is that they are bulkier than a plain bearing. This will mean that the engine big end for example will need a larger housing, making the connecting rod big end bulkier. This is not normally an issue with main bearings.

One interesting example of a four-cylinder engine made by Schillings uses a built-up crankshaft running in large-diameter ball races held in the crankcase, resulting in a very rigid set-up.

FLYWHEELS

Flywheels are used on stationary and marine engines to smooth the power output and to aid the balance of the engine by adding weight to the crankshaft assembly. It is important for the flywheel to be both properly balanced and also tightly secured to the crankshaft.

A flywheel is more important on a single-cylinder engine. With a multi-cylinder engine there are more power impulses per revolution, which provides smoother power output and enables slower tick-over speeds. From the point of view of smoothness, the bigger and heavier the flywheel the better. However, if the flywheel is too large, the engine will accelerate more slowly and, in the event of a failure, the energy in the flywheel may result in considerable damage to the engine.

Materials

Flywheels are normally made from steel or brass, to provide the necessary mass in a small diameter. However, on some very small engines, aluminium-alloy flywheels have been fitted and on slower-revving single-cylinder engines, spoked cast-iron flywheels are often used. On anything other than a slow-revving single-cylinder engine, a cast flywheel can end up out of balance if there is a blow hole inside the casting.

Machining Sequence

Whatever material is used, the wheel must be balanced and this can be ensured by using good-quality material and machining the majority of the outside faces and the shaft bore at one setting.

A typical disc flywheel, as fitted to a vertical single-cylinder engine, should be machined according to the following sequence:

1. Set the material to run truly in the chuck, face off the end and clean up as much of the outer diameter to size as possible, leaving it slightly oversize.
2. Turn down the starting pulley section (if required) and cut the groove, ensuring that two parallel areas are left for holding in the chuck.
3. Reverse in the chuck (the back face is now outwards) and face off to length and turn the rest of the outer diameter down.
4. Bore out any recess in the rear face before centring and drilling a pilot hole for the shaft.
5. Use a boring bar to bore out the shaft hole, which may be a taper, depending on the fixing. If it is tapered, the shaft or collet taper must be machined at the same setting.
6. The flywheel is now finished. Because the outer surfaces and bore have been finish-turned at the same setting, it will be balanced.

Shaft Fixings

The shaft fixing must hold the flywheel truly and rigidly to the shaft and also be able to absorb and possibly transmit the engine power impulses without moving on the shaft.

The first method, commonly used on slow-revving horizontal engines, is a plain shaft and hole, with a key to locate the flywheel. The key must be a good fit and the fit of the wheel to the shaft is also critical. Any slack will show up as a flywheel that runs out of true.

The second method is to turn a taper on the shaft with a corresponding taper in the flywheel, which is pulled on to the taper using a nut on a threaded portion of the shaft. This method is good but it is not always convenient to turn the shaft taper at the same time as the flywheel taper. It can also be difficult to ensure the correct location of the flywheel because a small reduction in the diameter of the taper will result in the flywheel pulling much further on to the shaft. A key may also be used for extra security, but should not be necessary if the shaft and wheel tapers are an accurate match. The taper is typically turned to 10 degrees included.

It is preferable to use a tapered hole in the flywheel with a matching split taper collet that closes on to a parallel section of the crankshaft and is located by a shoulder at the engine end. This means that the crankshaft is a simple parallel turning job and the collet can easily be machined at the same setting as the flywheel. The crankshaft hole should again be finished by boring and the taper turned without moving the job in the chuck. It is important that the hole in the collet is a push-fit on the shaft, with no slack.

The flywheel can be tried on the taper and the collet length can be set accurately from the back face of the flywheel. The front face of the collet should be just below the flywheel face when fully home.

Once turned to size and parted off, the collet is easily slit using a slitting saw in the mill; this may even be done carefully with a small hacksaw. Make sure that all burrs are removed from the edges of the cut.

Split-collet flywheel or propeller driver fixing. The collet must be a close fit on the shaft.

One point to make is that the collet is best made from a softer material than the shaft because then, if it does slip (as a result of a backfire, for example), it will not damage the shaft. High-tensile aluminium alloy is a good choice for collets and seems to cause no problems in use.

For those building aero engines, all the above comments apply equally to propeller drivers, which can use exactly the same methods. In this case it is the front propeller driving face which must run truly with the centre hole, so these two machining operations should be carried out at the same setting.

5   Connecting Rods

The connecting rod is one of the most highly stressed parts of an engine, having to take both the thrust of the power stroke and also the forces involved in reversing the direction of travel of the piston at each end of the stroke.

Two attributes of the connecting rod are particularly important: the distance between the big- and little-end centres and the parallelism of the big- and little-end bearings. The first is critical because any variation from the design length will affect the compression ratio. Any inaccuracy in the second will cause undue friction and wear on the bearings and possibly on the piston as well.

MATERIALS

The material used for connecting rods will depend on the type of engine being constructed.

For a slow-revving single-cylinder engine, steel is commonly used, with bronze bearings. This is mainly for fidelity to prototype.

For typical single- and multi-cylinder four-stroke and two-stroke engines, the almost universal choice is high-tensile aluminium alloy such as HE15. These connecting rods may or may not have bronze bearings inserted, because HE15 is a good bearing material when used with well-finished steel crankpins.

Cast-bronze connecting rods have also been used but have largely been superseded by aluminium alloy because of the lighter weight of the latter.

BIG ENDS AND SMALL ENDS

The small end of the connecting rod provides the location and bearing for the gudgeon pin.

Aluminium-alloy connecting rods can have a plain reamed small end running on a silver-steel gudgeon pin. Steel connecting rods will need a bronze bearing push-pressed in and reamed. Larger engines can also use needle-roller bearings running on a hardened gudgeon pin.

A lubrication hole must be drilled through from the top of the small end into the bearing for lubrication.

The big-end layout will vary depending on the type of engine. Slow speed horizontal engines will have bronze bearing caps bolted to the flat end of the (usually steel) connecting rod and provided with a lubrication cup.

A single-cylinder overhung crank engine can have a plain reamed bearing directly into an aluminium-alloy connecting rod or can be fitted with a reamed bush or even a needle-roller bearing.

Multi-cylinder engines or single-cylinder engines with a one-piece crankshaft typically have split big-end bearings, as this facilitates the assembly of the engine. These can run directly on the connecting rod or may have separate bronze bearing surfaces if desired. The big-end cap should be retained with steel bolts made from one of the tougher grades of mild steel. Steel socket screws are a good option here and, whatever is used, it is advisable to provide lock nuts or use wire locking or threadlock to prevent the possibility of things coming loose.

One other option is to use plain big ends in conjunction with a pressed-together or bolted crankshaft construction. This set-up was very common in early motorcycle engines but can be tricky to make accurately.

MACHINING CONNECTING RODS

Slow-Speed Horizontal Engines

Connecting rods for slow-speed open-crank engines are most often a round section and are turned from steel, with a taper from the big end to the small end. The sequence of turning is as follows:

1. Take a piece of suitable steel long enough to allow for a chucking piece on the big end.
2. The small end is often spherical and it helps to turn the ball end and to drill and ream the small-end bearing before turning the taper on the main part of the connecting rod. The best way to ensure accuracy is to set the blank up in a V block and then to drill and ream the bearing in the vertical mill. If a mill is not available, set the rod accurately in the vertical slide on the lathe for the process.
3. Once the small-end bearing is drilled, a centre should be put in the end of the small end. This is used when machining the tapered part of the rod.
4. The rod is then reversed in the chuck and the other end is faced and centred. The taper portion is then turned between centres with the tailstock offset for longer rods or with the top slide set over for shorter rods.
5. Ensure that the chucking piece is left parallel for finishing off.

If the big-end fixing is to be made integral with the rod, then a larger-diameter rod will be used and the taper portion will need to be turned down before cutting the taper. In this case, it may be easier to mill the rectangular bolting face to shape before parting off the chucking piece.

If the bolting face is to be added later, a small spigot should be left on the end of the rod to locate the bolting flange for silver-soldering when parting off. The flange can be made from rectangular-section mild steel.

High-Speed Engines

The connecting rods of high-speed engines can be machined from blanks of aluminium-alloy bar and this is probably best done in the vertical mill with a rotary table being used for finishing the outside shape. The machining sequence will vary slightly depending on whether the big end is split or not.

For a split big end, the sequence is as follows:

1. Face off one end of the bar and mark out the position of the big-end bolts on that

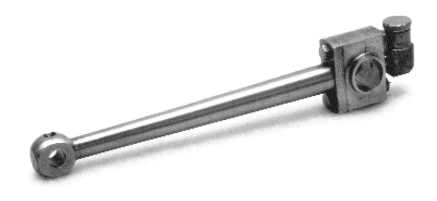

Steel connecting rod with split big end for a slow-revving horizontal engine. This type of connecting rod is typical of such engines.

Aluminium alloy connecting rod with needle-roller big end for a single-cylinder engine.

face. These are then drilled tapping size for the big-end bolts.

2. Cut off the big-end cap, ideally with a slitting saw or by careful hacksawing. After hacksawing, the mating faces will need cleaning up and the parts should be marked to ensure correct assembly later.

3. Drill out the bolt holes in the cap to the correct clearance size and tap those in the remainder of the rod for the bolts.

4. The two parts are then bolted together for the subsequent operations, which will also apply to a rod with plain big ends.

5. Clean up the wide faces of the blank in the mill before mounting in the machine vice to drill the big-end and small-end bearings. It is essential that the blank is parallel to the machine table, so check this carefully.

6. The big-end hole must centre on the split between the rod and cap. The bearings can be drilled and reamed or drilled and bored to size. If bearing bushes are to be used, they should be fitted at this stage and reamed to size if needed.

If the rods are for a multi-cylinder engine, it is worth making a simple jig to locate the two holes relative to each other if you do not have digital read-outs on the mill.

Once the bearings are bored, the outside of the rod can be shaped, and this is best carried out on a rotary table. It is advisable to make up a jig to locate the rod, so that it can be turned over and relocated accurately for machining both of the side and top faces.

It might be preferable to machine the middle portion of the rod first before machining the outline of the big and small ends. Make sure that the curve of the big and small ends merges cleanly into the middle portion of the rod. If any small ridges are created, use needle files to clean this away. If they are left, they may cause undue stress at that point. Connecting rods should have a good smooth finish all over.

Oil holes may be needed in the big and small ends and these can be drilled with the rod clamped between packing pieces in the milling vice.

If needle-roller big or small ends are used, they can be pressed in at this stage.

6 Cylinders and Cylinder Liners

The majority of four-stroke engines have separate cylinder liners in the cylinder block. The liner is either pressed in or may be removable and with water-cooled engines the liner may form part of the cooling space. This latter arrangement is what is known as a wet liner.

MATERIALS

Cylinder liners are made from either cast iron or one of the oil-hardening steels.

Steel liners are needed when the cylinder and cooling fins are in one piece and form a structural part of the engine. This is typical of radial and rotary engines (*see* Chater 20), and may also be used for higher-performance engines, in order to save weight, as well as in scale models of early aircraft engines.

Engines with aluminium-alloy cylinder blocks will have an inserted liner, which can be steel or cast iron. An engine with a cast-iron block can have the cylinders bored directly into the block. The inserted liner can be of cast iron or an oil-hardening steel such as EN30B. For the amateur, the easiest to use is cast iron, which wears well when used with cast-iron rings or pistons and requires no heat treatment, thus avoiding any problems with distortion.

In the past, case-hardened mild steel liners might have been suggested but, again, distortion may be a problem and this material therefore offers no advantage in these days of easy material availability.

MACHINING

A plain inserted cylinder liner is easily machined in the lathe from a length of suitable cast-iron bar. If good-quality cored cast iron is available it saves removing a lot of metal.

The machining sequence is as follows:

1. Chuck a length of bar with enough projecting to be able to machine the full length of the liner.
2. Machine the outside to overall size and then use large drills to remove the bulk of the metal from the inside.
3. Finish the inside with a boring bar and, when it is near to size, run the boring bar through at least twice at each setting to ensure that any spring in the bar is avoided.

Plain cast-iron cylinder liner.

4. If a steel liner is to be hardened, the bore should be left a fraction undersize to allow for honing, which will be needed after the hardening process.

5. Most liners have a small inside chamfer at the lower end to aid the insertion of the piston rings, so offset the boring bar and machine this next.

6. Now use a parting tool to machine off the liner, leaving it slightly longer than finished length to allow for cleaning up the top flange.

7. Reverse in the chuck and face off any top flange to length and diameter if this was not done at the first operation. This part of the process may be best done on a mandrel for long liners.

8. If desired, the liner can be honed (essential after any hardening) inside to remove any machining marks. If a sharp tool and slow feed was used for the boring, this should not be needed. You may like to use a small spring-loaded hone very lightly, with plenty of honing oil, just to polish off any marks.

For the amateur builder, it may be better to get steel liners hardened professionally – there is less chance of distortion if the process is carried out in a proper furnace that will heat the part evenly. Small one-off parts can often be put in with a batch of commercial parts at a reasonable cost – or even free, if you make friends with the heat treatment operator.

INTEGRAL COOLING FINS

If the liner has integral cooling fins, the cutting of these is best done before hardening with the liner on a mandrel after the bore has been machined as above. Use a sharp parting-off tool (with plenty of cutting fluid for steel) and retract the tool frequently to clear the swarf.

It is also sensible to use the lathe index to set the tool position accurately for each fin. Cooling fins are prominent on an air-cooled engine and nothing stands out more than uneven spacing or thickness of fins.

7 Pistons, Gudgeon Pins and Piston Rings

PISTONS

Materials

Pistons in miniature engines are almost always machined from aluminium alloy, either cast or bar stock. Cast pistons are typically LM51 with HE15 or HE30 bar stock; both are suitable. The usual exception is for small compression-ignition engines and for small-capacity engines, when cast iron is used.

Piston rings can be considered optional for engines with bores below about 15mm diameter but have been fitted to even smaller sizes. Rings are not usually fitted to cast-iron pistons, which can be made a close fit in the bore.

Castings for pistons will normally have a chucking piece on the crown to aid machining.

Machining

Preparing a Casting

There is one significant difference between machining a cast piston and machining one from bar stock. The first operation on the casting is to true up the chucking piece:

1. Hold the casting in the chuck with the chucking piece facing outwards and set the body of the piston to run as truly as possible.
2. Take light cuts on the chucking piece, to clean it up parallel.
3. Now reverse it in the chuck and check that the piston body is running as truly as possible and that the cored hole is also as true as possible.

From this point on, the machining sequence is the same as when using bar stock.

Machining the Piston Body

1. If using bar stock, set a piece of suitable bar stock in the chuck with enough protruding to machine the full length of the piston, and part it off.
2. Machine the outside of the piston to the finished size and face off the end.
3. Clean up the inside down to the gudgeon pin boss using a boring bar.
4. Drill the centre hole to slightly less than the dimension between the gudgeon pin bosses.
5. For larger pistons, an angled boring tool may be used to remove metal from between the gudgeon pin boss and the underside of the piston crown.
6. If rings are to be fitted, the ring grooves can be machined using a parting-off tool of the correct width. If the rings are available, then use a narrower parting tool and take two cuts to get the correct width to fit the ring. This should be a sliding fit with no slack and the groove depth should be 0.07mm deeper than the ring depth.

Milling the Inside

1. Mount the piston blank in the chuck upright on a rotary table in the mill and centre the piston under the tool.
2. Use an end mill to remove the rest of the surplus metal from the inside cavity and to finish the small end gap to the correct width, ensuring it is in the correct position. The correct position is usually central, but multi-cylinder engines have been built with the connecting rod offset in the cylinder, to provide clearance.
3. Next, for drilling and reaming the gudgeon pin hole, either mount the rotary table so that the piston is horizontal under the tool or transfer the piston to a V block. If possible, use the rotary table and drill from both sides before reaming right through. This helps to ensure that the hole stays central.
4. Return the piston to the lathe and part off from the bar/chucking piece, leaving a fraction on the crown for cleaning up.
5. Reverse in the chuck, gripping the piston lightly with soft packing to avoid damage, and face off the crown to the correct length.

GUDGEON PINS

The best material for gudgeon pins is silver steel. It can be hardened if desired, but experience shows that unhardened silver steel may be used without any problems at all.

The pin is parted off to the correct length and should be drilled out in the centre to about one-third of the outside diameter, for lightness.

Unless the engine design provides for fixing the pin in place, brass end pads should be fitted to the ends, so that the ends of the pin do not score the cylinder bore. These can be a press-fit or held in place with Loctite or similar.

PISTON RINGS

The making and fitting of piston rings in miniature engines is a wide-ranging subject, but the intention here is simply to detail a method of making plain piston rings easily and quickly, which can be carried out in any model engineer's workshop.

Piston rings are used to provide good piston to cylinder seal, particularly in engines with

Aluminium alloy piston with two cast-iron piston rings.

Silver-steel gudgeon pin with brass end pads. The pads prevent damage to the cylinder wall in use.

aluminium-alloy pistons running in cast-iron or steel bores. These engines need the extra sealing provided by properly fitted rings because there has to be more clearance between the piston and cylinder, to allow for the different expansion rates of the piston and liner. Engines with plain aluminium pistons can be very difficult to start from cold because of the bad seal that exists before they warm up.

Engines with a bore of less than around 15mm can be made without rings but, with anything larger, rings will be of benefit.

Materials

The most common material used for miniature piston rings is cast iron. This needs to be good-quality fine-grained material and, if cast-iron cylinder liners are used, a piece of the same material is best for making the rings.

Once machined, the rings have to be 'gapped' and heat-treated to provide the spring that gives the close fit in the bore.

Dimensions

Prolonged discussion on the subject of piston-ring dimensions has led to a set of generally accepted criteria based on the bore size. However, it should be emphasized that these criteria are guidelines only, and dimensions in individual engine designs may depart from them.

With a bore diameter of 'D', the radial thickness of the ring is best set in the range D/20 to D/30. The depth of the ring is normally the same dimension – in other words, a square cross-section – but this is not critical. One of my engines has rings that are rectangular, with the depth being twice the radial thickness, simply because the rings were available. This engine has run well for many years and does not burn oil at all.

All piston rings are split and must be installed with a small gap between the ends when in the bore. This gap should be between D*.002 or D*.003 inches. The gap is preset to a larger dimension during the heat-treatment process so that the ring has to be compressed to fit in the bore. A good guide for the preset gap is D/10, which has given good results in the engines.

The above are guidelines, and there are examples of rings that are considerably outside these parameters, but which work well none the less. One example is a $^{13}\!/_{16}$in diameter ring with $^{1}\!/_{16}$in cross-section equating to D/13.

The problem with thicker rings is that they may be too stiff and exert too much pressure on the cylinder wall. They will also be more difficult to fit and may be more prone to breaking during that process. The ring should be a close but sliding fit in the groove and the radial groove depth should be approximately 0.003in greater than the ring thickness.

Making Piston Rings

The manufacture of a set of piston rings is a process involving machining operations followed by heat treatment.

Turning the Stock Bar
1. Chuck a suitable length of close-grained cast-iron bar and turn the outside diameter to a very tight fit in the cylinder bore. If the fit is too slack, the resulting ring gap will end up too large. It is better to have the rings slightly large rather than too small.
2. If you are going to use a slitting saw to split the rings, make the outside diameter greater than the bore by an amount equal to one-third of the slitting-saw thickness.
3. The surface finish should be the best possible. Some constructors lap the outside, but this is not necessary.
4. Bore the inside out to leave the correct radial thickness of the finished ring.

Parting Off the Ring Blanks
1. Part off the individual rings from the bar using the machine indexes, to obtain the correct thickness of ring. If the piston is available, the ring can be checked for fit against it.
2. Once the correct fit is established, part off several rings at this setting, allowing for some spares in case of breakage later when fitting.

A group of piston-ring blanks ready for breaking and heat treatment.

3. It is sensible to use a file at an angle in the groove to remove the burrs before parting off fully. Also, once each ring has been parted off, clean the burr off the inside of the bar before cutting the next ring. This is easier to do at this stage.
4. Next, remove any burrs from the rings and give each a light rub with the ring flat on a piece of very fine emery on the surface table, or on the surface of a wide diamond file.

Breaking the Rings
Breaking the ring is easily carried out in one of three ways:

1. If you have a very fine slitting saw, this can be used carefully with the ring held in the milling vice. Remember that you must allow for the saw thickness when turning the rings. Some constructors cut the rings at an angle, but this should not be necessary with properly fitted rings.
2. The second way is to file a small nick in the inside of the ring with a triangular needle file and then to press the ring down a taper mandrel (a Morse taper is often used) of suitable size. The ring should break cleanly at the nick, but this may result in the ring ends being uneven.

3. The third way is to use a very sharp cold chisel to break the ring, with the ring flat on a surface plate or similar. For thin rings, a small knife blade may be used.

Using a taper has always produced good results, with no breakages other than where they are wanted.

Heat Treatment
The heat treatment of rings has been the subject of much discussion over the years. The following method is easy to carry out in the home workshop, and works well.

The heat treatment process does two things. First, it changes the characteristics of the metal structure, to give the necessary spring to the ring. Second, it also presets the gap in the ring so that the ring is under compression when in the cylinder.

The first essential is to use a jig to clamp the rings firmly during the process, to avoid distortion and maintain the correct preset gap. This jig is a simple turning job and consists of a short piece of mild-steel bar stock, turned with a spigot that is the inside diameter of the rings when set to the correct preset gap. The depth of the spigot should be slightly less than the total thickness of the batch of rings being treated.

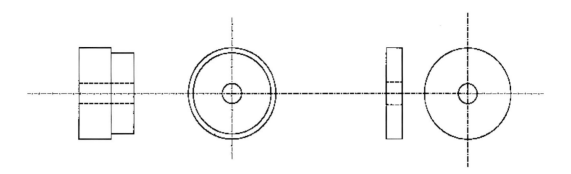

Piston-ring heat-treatment jig.

No more than five or six rings should be done at each session.

The rings are clamped with a thick steel 'washer', which must be thick enough not to distort when everything is at red heat.

The batch of rings is clamped in the jig and the gap is checked before finally clamping up. If the gap is too small, a small piece of bar stock can be inserted between the ring ends to maintain the gap. This can be left in during the heat-treatment process.

The whole assembly is then heated up in the brazing hearth to red heat and held at that temperature for at least ten minutes before being allowed to cool naturally.

Once cool, the rings can be removed and given a light rub with fine emery to remove any scale. Each ring is then tried in its cylinder and the final gap is checked. If necessary it can be increased by filing the ring ends with a fine needle file.

The heat-treatment jig in use with a batch of rings.

8 Crankcases

The production methods used for the crankcase will vary greatly depending on whether the crankcase is machined from a casting or from solid. The method chosen will depend to some extent on the equipment available and also on the availability of castings.

For those with small machines, machining from the solid may be the best option because, in such designs, the crankcase is often made up of smaller items bolted together, rather than one large casting. For example, a single-cylinder engine would typically have a crankcase body with bolted-on front bearing housing and separate cylinder block. Each of these could be machined on a smaller machine than the equivalent complete casting.

Single-cylinder, open-crank horizontal engines differ in that they are generally built up of separate castings for the base, the main bed plate, the cylinder and the cylinder head. The main bed plate is the equivalent of the crankcase.

MATERIALS

For open-crank engines the bed plate and base are usually castings, either aluminium (LM4) or cast iron. There is no reason why such engines cannot be machined from the solid but the complex shapes and size of some parts make this more difficult.

Single- or multi-cylinder enclosed engines may be machined from the solid or, again, from castings. The most common material is aluminium alloy but cast iron may be used if the cylinders are to be bored directly in the block.

Crankcases may also be fabricated from steel (or, more rarely, brass) by welding or brazing, in which case the resulting fabrication will be finish machined in the same way as for castings.

MACHINING

The main difference between machining from the solid and machining from castings is in the amount of metal to be removed. Obviously, with a casting, the centre of the crankcase casting may be cored to leave the inner space requiring little or no machining. For smaller engines, the casting may not be cored, in which case the process is very close to that used when machining from the solid.

Machining from Castings

When machining from castings, it is essential to carry out some preparatory work before commencing the machining proper.

With any casting, the first thing to do is to measure up the casting and check the various dimensions to ensure that, when the different surfaces are machined, no areas of the casting are left with too little metal and that all the various holes are centred. For example, if the inner void in the crankcase is cast slightly off centre, then this must be allowed for when setting up, to avoid one side of the crankcase being left thinner than the other. It is not unknown for castings to be supplied that are incapable of being machined.

Once the casting has been checked, the next stage is normally to clean up the casting to remove any flash and blemishes, which may cause problems when setting the casting for machining. This typically includes small projections left from the casting vent holes, which are on otherwise flat surfaces.

The next stage is to clean up the datum surface from which other dimensions will be taken. This surface will often be indicated on the drawing or be obvious from the dimensioning. This cleaning up can be done in the lathe or mill or sometimes carefully by hand. If using machines, make sure that the casting can be held firmly because the first cuts on castings are often intermittent and, particularly with cast iron, may need to be fairly heavy to get below the skin of the casting.

Some smaller engine castings may be supplied with a cast spigot for holding the work during the machining process. In this case, this should be cleaned up first, to provide a good accurate gripping surface. It may also be necessary to clean up another face to provide a decent clamping surface. For example, if machining a four-cylinder engine crankcase where the cylinder head face is the datum, it is often best to clean up the bottom face with a light cut to provide a flat face for clamping the casting when machining the top face.

For single-cylinder engines with the main bearing housing integral with the crankcase, the inside of the crankcase is often machined first with the casting held in the lathe by the front bearing housing. In this situation it is sensible to bore the main bearing housing at the same setting.

The casting will then be set up using the inside as the datum, often on a simple jig for further machining operations.

One of the problems with castings is locating the correct datum points when marking out. This can be overcome by inserting temporary plugs in order to be able to mark out centres of cored-out holes. A typical example is when locating the cylinder centre line in a single-cylinder engine. If an aluminium or hard wooden plug is inserted into the cored hole,

Small crankcase casting set up for machining in the four-jaw chuck. Care must be taken not to distort the casting, and bridging pieces may be needed across the cylinder face, as shown.

the centre can be marked on the plug and this centre can be used to set up the casting for boring out the cylinder recess.

The other problem with machining from castings is holding the casting securely in the lathe or mill. Simple jigs are often used for this and may well save time in the long run. If the circular inside of a single-cylinder crankcase is machined first, then a mandrel can be turned to be a close fit in this and the mandrel used to mount the casting securely for further operations.

It is sensible to think through all the machining operations and decide on any fixtures that might be beneficial *before* starting machining.

Once the faces have been machined, the various bolt holes and other small apertures are machined in the mill or lathe, using the

Milling crankcase datum surfaces using a wide cutter in the mill. This could also be done using the four-jaw chuck in the lathe.

machined surfaces as the datum for setting up and marking out.

Because each casting and engine is different, it is difficult to give a defined sequence for machining. However, the above hints should help you to work out the best sequence for any particular engine.

Machining from the Solid

Machining from the solid is often easier than from castings because it means starting off with a rectangular or round piece of stock that can be held easily in the lathe or mill. One point to make is that a holding allowance may be needed in order that the crankcase can be machined easily. This is particularly true of small single-cylinder engines.

When machining crankcases from solid, start by cleaning up the main datum surfaces in order to provide accurate surfaces for marking out or setting up. As an example, for a single-cylinder crankcase with bolted-on cylinder block and main bearing housing, first machine the two bolting faces accurately at right-angles to each other. All other machining can then be carried out with reference to these two surfaces.

Once the datum surfaces have been cleaned up, the block can be machined to final outside dimensions before moving on to the inside.

The surplus metal inside the crankcase needs to be removed – it is often easier to bore the large holes for the cylinder location and the front bearing location through the block first, because this removes a lot of the metal from the inside.

For a single-cylinder engine, set up the block in the four-jaw chuck and bore the holes with a rigid boring bar. You might prefer to use the boring head in the mill, but on the lathe you can use the power feed, which you may not have on your mill.

Once these two holes have been bored, the rest of the metal can be removed from the inside using the mill. Drill most of the remaining metal away before using end mills to clean up the inside to size. You will need long

Boring the front bearing location in a crankcase machined from solid using the four-jaw chuck in the lathe. This operation could also be done in the milling machine using a boring head.

series end mills here; make sure they are sharp and take small cuts, to obtain a good finish. At this point another benefit of boring the large holes first becomes apparent – the swarf produced from the inside can escape through the large holes in the block. For this reason, it pays to set the block up on parallels.

Once the inside is cleaned up, then, as with castings, all the other bolt holes and so on can be machined. This procedure may be followed by any fancy shaping of the crankcase external shape – either to suit the whim of the constructor or because the engine is a scale model of a full-size prototype – which can be carried out with a mixture of machining and hand filing and polishing.

MULTI-CYLINDER ENGINE CRANKCASES

There are some important points to note when machining crankcases for multi-cylinder engines.

Because of the need to ensure the correct relationship of the cylinders to each other and the crankshaft centre line, fixing datum surfaces

Milling out the inside cavity of a crankcase in the milling machine.

is important. It is possible to use one crankcase end, the cylinder-block bolting face and one side edge of the block as the datum. These should be cleaned up so that they provide an accurate location for marking out or setting up.

With multi-cylinder engines, the cylinder block and crankcase are often integral, in which case the cylinder-head bolting face should be one of the datum faces. In this case, the cylinder locations will also need to be bored in

the correct relationship to each other. This can be done in the mill using the indexes to set the correct distance or, if the lathe is used, the cross-slide index can be used.

In one of his articles, Edgar Westbury describes using a face plate with a locating bar and a spacer for indexing the correct distance. However, it is generally easier to clamp the crankcase on its side on the cross slide and use a boring bar in the chuck.

Use of Jigs for Machining

The use of jigs can make the machining of complex shapes easier.

When machining from solid, the use of jigs will generally be limited to simple drilling jigs for ensuring the correct location of bolt holes in relation to each other, or to other parts. In such cases, the jig can often be made to locate in a previously machined hole or against a machined surface. An example is a jig made to drill the cylinder bolt holes in a crankcase, which has been drilled at the same setting as the cylinder

and has a spigot to locate in the cylinder hole in the crankcase.

Jigs may be essential in order to facilitate holding a casting for some machining operations. This is particularly true when machining such things as angled faces that have to match for bolting to other parts.

Generally, if a part has to be removed from a machine and then replaced in the same position for another operation, a jig should be considered.

With multi-cylinder engines, jigs are more important because they will be used to ensure that such things as the cylinder bolt-hole patterns are identical.

Simple jigs may also be used to locate two parts together in the correct relationship so that fixing holes can be drilled through both together. An example of this is a timing case cover, which has to be in accurate alignment with the crankshaft so that the bearing in the housing is in correct alignment. In this case, making a simple turned mandrel that locates in the crankshaft bearing and the cover bearing will ensure correct alignment when drilling the fixing holes.

Simple cylinder bolt-hole drilling jig with spigot to locate jig in crankcase cylinder register.

Many designers will show details of jigs on drawings or in any accompanying notes but, even if this is not the case, thought should be given to jigs at an early stage.

9   Cylinder Blocks

The cylinder block can be one of several types depending on the method of cooling used, whether separate cylinder liners are used, and whether the cylinder block is integral with the crankcase.

If the block is integral with the crankcase, then the comments about crankcases above apply.

If the block is a separate part, it may be of a different material and will probably be machined separately. In some engines, however, it may be beneficial to part-machine both the crankcase and the cylinder block before finishing the machining with the two fixed together. It may be beneficial on a multi-cylinder engine to bore the block and crankcase together to fit the cylinder liners in order to ensure that all the bores line up correctly.

MATERIALS

Materials used for cylinder blocks are the same as for crankcases but do not have to be the same as the crankcase on any particular engine. Constructors may use an aluminium crankcase with a cast-iron block with the cylinder bores machined directly in the block, rather than have the added complexity of separate cylinder liners.

MACHINING CYLINDER BLOCKS

The machining of cylinder blocks falls into two categories: single-cylinder engines and multi-cylinder engines.

Single-Cylinder Engines

The single-cylinder engine block (often referred to as a cylinder jacket) is most often a turning job followed by the drilling of the various fixing bolt holes. There are some designs with square or rectangular jackets and these will be treated in the same way as a multi-cylinder engine.

If a casting is used, it should be set up in the lathe chuck to run as true as possible and faced off true. If the casting has been cored out for the cylinder bore, the true running of the core must also be checked. It will probably need a compromise to get both running within acceptable limits, the object being to ensure that both the bore and the outer dimension can be finished to the correct size.

With a jacket turned from stock bar, a suitable length of bar should be set up. However, beware of large overhangs from the chuck if attempting to have enough protruding to machine the whole of the outside at one setting. It is often better to have the correct length of bar and to reverse it in the chuck to finish the outside, or to mount the jacket on a rigid mandrel set between centres to work on the outside after boring the centre.

The centre hole for the cylinder liner is best bored using a rigid boring tool although, on small engines, the bore could be drilled and reamed to size. The liner fit will depend on whether it is intended to be removable, to enable the builder to swap the engine from an air-cooled to a water-cooled jacket. In this case, the liner needs to be a slide fit in the jacket.

For a more permanent fit, liners can be

pressed in or made a shrink fit. In this case, the standard tolerances for the preferred method will be used.

Loctite should not be used for fitting liners in air-cooled jackets. This is because the film of adhesive forms an insulating barrier between the liner and the jacket, thus inhibiting the transfer of heat between the two. Loctite may, however, be used for 'wet' liners in water-cooled heads because the heat transfer to the cooling water is not affected.

A water-cooled liner will also need the water space boring out to a larger diameter. This is carried out with a boring tool with an offset tip large enough to reach the correct diameter without the bar fouling on the smaller bore. Another option is to turn the outside with two thick flanges and then to Loctite a length of tube over these to form the water space. This results in a dry liner because the water is totally contained within the cylinder jacket.

Once the centre bore is machined, the outside of the liner can be tackled. For a water-cooled jacket, this may be no more than finishing to size and turning any fixing flanges on the crankcase end. Air-cooled jackets will

require fins to be machined and the same comments made above in relation to integral fins apply.

Great care is needed when cutting thin fins in aluminium and it is essential to use a sharp tool and plenty of cutting fluid. For aluminium alloys, use a mixture of paraffin and neat cutting oil. It also helps to use a slightly narrower tool than the space between the fins and then to move the tool slightly to each side to cut the final size gap each time. Effectively this is facing the surface of each fin with the tool edge but it does help to avoid swarf getting trapped between the tool and the fin surface. This can cause two problems: the tool can get broken and the swarf can push the narrow tool off centre, resulting in a fin that is thinner than it should be.

The final operations will be drilling all the various fixing holes, some of which will not be done until the mating part has been made.

The above process will also apply to multi-cylinder engines where the cylinders are separate items, such as early aero engines. In this case, the tops of the cylinders are often skimmed to length in the mill, with the jackets

Boring a cast-iron cylinder liner. A slow speed is needed for this and the boring bar should extend for the minimum needed to avoid chatter.

bolted to the crankcase or a jig to ensure that they are all the same. This is most often done with engines that have separate cylinders with a one-piece cylinder head.

Multi-Cylinder Engines

Blocks for multi-cylinder engines are most often machined in the mill or by milling in the lathe. The machining sequence for the block will be more or less the same for castings as for those machined from solid parts. The difference is that, in castings, the cylinder bores and possibly some of the water passages will be cast in and so less metal will need to be removed. If a casting is used and the bores are cast in, it is preferable to machine the bores at an early stage.

The sequence for machining a casting is as follows:

1. Check it over and ensure that any cast-in cavities are in the correct place in relation to the outside. If the cylinder bore cores are slightly offset it may be necessary to bore the cylinders offset from these holes, so make sure there is enough metal to carry this out.
2. Clean slag and flash from the casting, then clean up the top and bottom surfaces to give the correct cylinder block height.
3. Cut the cylinder bores using a micrometer boring tool, trying to ensure that they are as close as possible in size to each other. The spacing between the bores is also important and for those with digital read-outs this is easy to achieve. For those without, use the machine indexes, with marking out as a final option.

Four-cylinder engine cylinder block set up in the mill for boring with a boring head.

4. Once the bores are cut, the outside surfaces can be machined with reference to these and then the fixing and cooling holes can be drilled. Things such as cylinder-head fixing holes are often left until the head itself is finished and the holes are then drilled using the head as a template.
5. Fixing points for ancillary items such as water pumps will also be finished once the relevant item is complete.

10　Cylinder Heads

The cylinder head of any four-stroke engine will vary, from being very basic in the case of a side-valve engine to being a very complex piece of machining for an overhead-camshaft engine. The side-valve head may not be much more than a flat plate with the combustion chambers machined and spark-plug holes in it. Any overhead-valve engine will have a complex head, which has to house the valves, spark plugs, inlet and exhaust ports and also all the various bolt holes for fixing the head and parts such as the rocker supports or camshaft housings.

MATERIALS

The materials most commonly used for cylinder heads are cast iron or aluminium alloy, in either cast or bar form.

For single-cylinder horizontal engines, cast iron or cast aluminium may both be used. The advantage of aluminium is that it is easier and cleaner to machine. In addition, for model aircraft use, the fact that it is lighter will be advantageous.

Marine or stationary engines often use cast iron, which removes any possible need for separate valve seats, although, with the correct choice of aluminium alloy and valve material, this may not be a factor. Cast-aluminium heads are typically in LM4 grade and will need inserted valve seats. Heads machined from bar stock can be HE30 or, if the valves are to seat directly in the head, HE15 is the sensible choice.

COOLING

The method of cooling used has a significant bearing on the design of the cylinder head. An air-cooled engine will be provided with cooling fins, which in current engine designs may be in-line vertical fins on the top of the head, often supplemented by horizontal fins on the edges. Some engine designs have just the edge fins.

A water-cooled engine may have an air- or water-cooled head depending on the designer's choice.

If the head is water-cooled, provision must be made for cooling passages in the head. These may be nothing more than drilled passages, linked to the cylinder block, but in more sophisticated designs the head may be made with cast-in cooling passages often closed by a separate cover to make manufacture easier. The Edgar Westbury *Seal* design has cast-in passages with a flat cover plate bolted on to the top under the cylinder-head bolts.

An air-cooled cylinder head on a water-cooled engine will be of similar design to that on an air-cooled engine, but may be with or without cooling fins.

MACHINING

Because the design and layout of individual cylinder heads varies considerably between different engines, it is possible here only to give general guidelines relating to the sequence involved.

Top and underside views of an overhead-camshaft cylinder head machined from the solid.

Single-Cylinder Heads

For single-cylinder heads, many of the machining operations may be carried out in the lathe. The following sequence also applies to multi-cylinder engines with separate circular heads. Such engines may benefit from the use of jigs, to ensure consistent machining for each head.

Cylinder Face and Combustion Chamber

1. Set up the bar stock or casting in the chuck and machine the cylinder bolting face and combustion chamber first.
2. Hemispherical heads will need to be turned with a spherical turning tool or, for very small engines, a specially ground form tool.
3. If bar stock is used it should be possible to have a piece long enough to be able to machine most of the outside diameter of the head as well. It is useful to mark the centre point of the head lightly, as an aid to setting up later on.
4. If a casting with a holding stub is being used, the position of this may dictate the machining sequence.
5. Once the cylinder face is machined, the outside of the head can be turned to size and any fins cut in the circumference.
6. The head may then be parted from the bar stock and reversed in the chuck for turning the top face if applicable.
7. Cast heads with fixing stubs cast on for rockers or other parts will have to be machined in the mill or by using the vertical slide in the lathe.

Boring the Valve Seats and Chambers

If the valves are in a flat head and are not angled, this job is best carried out with the head set cylinder face up in the rotary table. Drilling and reaming or boring the valve passages and valve guide locations should be carried out at one setting, to ensure that everything is concentric.

With one-piece inserted guides and seats, these are pressed in complete and so at this stage all that is needed is to drill and ream suitable holes in the head.

If the valves have the seats cut in the head, care must be taken to cut these correctly at 45 degrees with a suitable boring tool. Do not rely on a centre drill or countersink for this.

Boring valve chambers before cutting the valve seats.

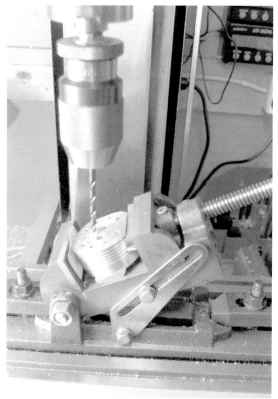

Drilling the spark-plug hole with the head mounted in an angle vice on the mill.

At this stage, with the job on the rotary table, it is sensible to drill the head bolting holes, ensuring that they are in the correct relationship to the valve chambers.

Heads with angled valves will need more complex setting up and this may mean making a special jig.

Drilling the Spark-Plug Hole

The spark-plug hole is most likely to be at an angle in the head and so it is essential to set the head at the correct angle on an angle plate or angle vice. One method is to drill a pilot hole at the correct angle from the combustion chamber outwards so that the hole is in the correct location in the inside of the head, which is likely to be more critical than the exact position on the outside. After doing this, you need to reverse the head on the table and pick up the pilot hole before enlarging it to the correct tapping size for the plug. The standard sizes are ¼in × 32tpi for glow plugs and smaller spark plugs and 10mm × 1mm pitch for the larger sizes. (Note that the 10mm spark-plug thread is a finer thread than the standard metric threads.)

Once the hole is drilled tapping size, it must be counter-bored to clear the spark-plug body and allow the plug electrodes just to enter the combustion chamber. If you have a counter bore of the right size, it may be used; if not, use a slot drill slightly smaller than the finished size and go slightly less then the correct depth, before finishing with the correct size. It is important to leave a flat surface for the plug to seat on at the bottom of the hole.

Once this is done, the hole can be threaded using a tap in the mill or drill at the same setting to ensure that the thread is truly perpendicular to the plug seating.

Drilling the Inlet and Exhaust Passages

Next, the inlet and exhaust passages have to be drilled from the correct locations on the outside to meet the valve chambers. If pressed-in one-piece seats and guides are used, these should be inserted first. The drilling is best done in the milling vice and may involve setting the head at an angle. This may be judged by eye with a drill in the chuck or by using a protractor, if the angles are defined on the drawing.

If there are flat faces for bolting the exhaust and inlet manifolds too, these can be milled first.

When drilling the passages, start with a small drill first to check the alignment, and enlarge the hole in stages.

It helps when drilling holes at an angle on flat surfaces to use a small slot drill to make a flat at right-angles to the drill. This will help to stop the drill wandering.

Cutting the Fins

Any fins on the top of the head can now be cut. A slitting saw is preferable for these because they may be quite deep and small-diameter end mills may move off line.

Set the head up on edge with the slitting saw in the vertical mill, cut the fins on one half of the head and then turn it over for the other half. If you have access to a horizontal mill, all the fins can be cut at one setting, although clamps may need to be moved around during the process to allow access to all the fins.

Use the mill indexes or digital read-outs to ensure correct and even spacing of the fins.

Finishing Off

Once the fins are cut, all that remains is to drill any remaining fixing holes and, if the head has an integral cam box, to mill this to size.

Any fixing bosses for rockers or other items must also be machined, which is normally a milling and drilling operation. Single-cylinder

Cutting the fins in an aluminium alloy cylinder jacket using a mandrel in the lathe.

LEFT: Gas engine cylinder head showing integral fixing bosses and brackets, making it difficult to hold for machining.

BELOW: A model based on the Volkswagen flat four engine built by Brian Perkins for use in a large model aircraft. This view shows the one piece cylinder heads for each pair of cylinders and the vertically split crankcase.

horizontal engines often have a multitude of bosses and brackets on the head for valve-gear rockers and camshaft supports. These can be difficult to mark out using normal methods and it is often better to use the machine indexes in relation to the bore centre or other suitable point.

As with other parts, some of the fixing holes may be left until the mating part is made.

Multi-Cylinder Engines

Cylinder heads for multi-cylinder engines will need more work in the milling machine or vertical slide, although, if the combustion chambers are round, they can be turned with the head on a face plate or, for smaller engines, in the four-jaw chuck. For the majority of engines, milling will be the preferred option.

Because most such heads are basically rectangular in shape, the first operation is to clean up the top and bottom faces in the mill or lathe, and to machine the head to the correct height.

The edges of the head are next; at this stage they should just be cleaned up to size. Once this has been done, the combustion chambers can be milled to the correct outline using the lathe indexes or, if not, to a marked-out outline. If the chambers are circular, a boring head can be used to ensure truly circular edges, after milling most of the metal away. Often, the edges of the chambers are rounded, in which case a ball-nose end mill can be used.

For overhead-valve engines, the valve chambers can be cut as for the single-cylinder head.

It is also sensible to drill as many bolt holes as possible, ensuring their correct relationship to the combustion chambers.

The exhaust and inlet passages are then drilled with the head set on edge in the milling vice.

A fine multi-cylinder engine, the 200cc capacity supercharged V8 by Ingvar Dahlberg.

11   Valve Gear

The valves and associated items (guides, rockers and tappets) are critical parts of any four-stroke engine and require care in manufacture and installation if they are to function correctly.

In the majority of four-stroke engines, the valves are mushroom-shaped poppet versions. The opening and closing of the valves is controlled by the rotation of the cams, as is the valve lift.

One exception to this is with the spring-loaded automatic inlet valve fitted to some horizontal engines. In this case, the exhaust valve is controlled from a cam and the inlet valve is held shut by a light spring and opened by atmospheric pressure when the pressure in the cylinder is reduced as the piston descends on the inlet stroke.

The design of the valve gear is affected by, among other things, the rpm and power range of the engine. In a slow-speed engine, the valves operate at a slow speed and so the inertia of the valves becomes less important, meaning that they can be heavier without adversely affecting the performance of the engine. This does not mean that weight should be ignored; all valve gear should be as light as possible without compromising strength.

In high-revving engines, the valves are operating at much higher speeds and excess weight can cause problems, resulting in the valves not closing when they should, often because the tappets are not following the cam contours exactly.

One of the difficulties encountered when making valve gear is that the parts are often very small and need a high degree of accuracy in manufacture.

VALVES

The valves are the key part of the valve gear; all the other parts of the valve gear are there to ensure that the valves open and close at the correct times.

Material
In the past, constructors used a variety of exotic materials for valves, but today the majority of amateurs use free-cutting stainless steel. This material is easy to machine and provides long life in the typical modern engine. Model engines generally do not run as hot as the full-size versions and so, unless an extremely high-performance engine is being considered, there is no reason to specify any other material.

Machining
Machining valves is a turning job but is complicated because the stem is typically around 3 or 4mm in diameter for a 10 or 15cc engine, and is long in proportion to that diameter. This means that the stem will need to be supported with a centre during the turning process. It is also important that the stem and seat are concentric so that the valve seats correctly and does not leak.

The suggested sequence for machining valves is as follows:

1. Chuck a piece of stainless steel of suitable diameter in the lathe and centre the end, so that a small centre can be used for support in the turning process. An alternative to the use of a centre is to use a tool-post-mounted small-

diameter turning tool.

2. Extend the material from the chuck, leaving enough protruding to be able to turn the complete stem, head diameter and seat at one setting. (The stem will usually need to be longer than required in order to accommodate the centre. For very thin stems, a conical female centre can be used; the stem might need to be made even longer to allow for this.)

3. Bring up the centre for support and then turn down the stem in easy stages until it is close to the correct diameter.

4. Take very light cuts to bring the stem to the correct diameter. During this operation, make several passes with the tool at each setting, to compensate for the spring in the stem. Check the diameter at various points along the stem to ensure that it is parallel. The objective is to achieve a close-running fit of the valve in the guide, which will typically be reamed to size.

5. Some constructors use an external hone on the stem to obtain a good finish and to ensure that it is parallel, but this should not be necessary if a good sharp tool has been used. The junction of the stem and head should be of a sufficiently large radius to provide maximum strength at this point. This also helps the gas flow through the valve.

6. The seat is the next part to be turned and this should be done with the tool post offset, to give a 45-degree angle to the seat. Do not be

Machining the stem of a valve using a small rotating steady in the tailstock.

tempted to plunge cut using a straight tool set at an angle; you will not get a good seat using this approach. The finish must be as fine as possible because the valves are not normally ground in as they are in full-size engines.

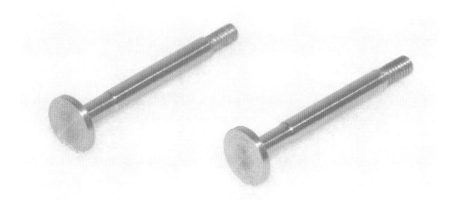

Completed valves with threaded ends for tappet.

Once the seat is finished, the complete valve can be parted off from the stock. If your parting tool leaves a good finish, you can part to exact length. If not, you should allow for a very light skim later. If you do need to skim the valve head off, use a split bush in the lathe to hold the valve by the stem as close as possible to the head and take very light cuts, being careful to avoid bending the valve.

The last operation is to cut the groove or thread for the valve-spring retainer. In either case, use a split bush in the chuck to hold the stem with just enough protruding to do the job.

If a groove is used, this is likely to be very small, and a special tool will need to be ground to cut it. Lengths of broken hacksaw blade can be useful to make this sort of tool.

On an overhead-valve engine, with the tappets acting directly on the cam, the valve-stem length may need to be set once the engine is assembled. Again, this will be done using the split bush in the lathe.

VALVE GUIDES AND SEATS

These two parts are considered together because, in many engines, they are made in one piece. The one-piece design was used by Edgar Westbury for many of his engines and has the advantage that if the seat and guide are machined at one setting then they will be certain to be concentric with each other.

Materials

There are two main choices for valve guides: phosphor bronze or cast iron. Both provide good wear qualities when used with stainless-steel valves and have been used widely. The same choice applies for one-piece valves, guides and seats.

If separate pressed-in valve seats are used,

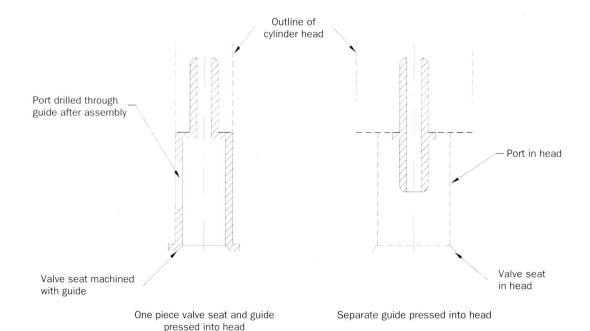

Two types of valve guide: (left) one-piece seat and guide, and (right) separate guide with valve seat in head. The separate guide provides greater bearing area for the valve stem but the one-piece version is easier to make accurately.

cast iron is probably the best material and the seat should be cut after it is fitted, to avoid distortion. This type of arrangement is not common on model engines.

Valves can also seat directly in the cylinder head. This is obviously true of engines with cast-iron heads but can also be done with aluminium-alloy heads if one of the tougher grades such as HE15 is used. In this case, the valve stems can also run directly in the head, although inserted guides may be preferable because they can be extended down the stem (into the valve passage), to give more support at the lower end and a longer bearing surface.

Machining

Machining valve seats and guides of either type is a basic turning operation but the sequence is important:

1. With the one-piece arrangement, the hole for the stem, the chamber and seat must be machined at one setting. Because this type of guide often has a flange at the seat end, this is easiest if the outside is machined before the part is reversed in the collet chuck, or a split bush in the chuck brackets, and the inside surfaces are machined.
2. Drill through for the valve guide hole before drilling and boring the valve chamber and cutting the seat.
3. To ensure things are concentric, for a plain valve guide, the part that is pressed into the head and the guide hole should be machined at the same setting. It is advisable to press the guide into the head using a special tool in the press or drill, to ensure the guide stays straight. It may also be necessary to run the reamer through after pressing, in case the hole has compressed slightly. If this is the case, it should be done in the mill or drill and not by hand.
4. With separate valve seats, the outside and the seat should be machined at one setting, with a final cut for the seat after the seat is fitted.

VALVE SPRINGS AND RETAINERS

Valve Springs

The majority of four-stroke engines use plain compression springs to close the valves but some early engines use hairpin or volute springs.

For those making their own compression-type valve springs, the best material to use is piano wire. The spring should be wound on a mandrel in the lathe; the mandrel diameter will need to be smaller than the finished inside diameter of the spring. The difference will have to be found by trial and error. If possible, the wire should be fed in through a friction device, to apply the correct tension. This should be mounted on the top slide and the power feed used to provide the correct advance between spring turns.

The ends of the spring must be ground off flat, to provide a good seating face and also to reduce local stresses in use.

For the majority of engines, suitable springs can be purchased very cheaply and this is the easiest way of obtaining good-quality springs.

If the engine design uses what are known as 'hairpin' valve springs, these will probably have to be made specially. Again, piano wire is the best material and, if several springs are needed, suitable jigs will help to achieve consistent results.

Volute springs are made from flat spring steel and are arranged so that, when they compress, the turns telescope into each other. It is possible to make them, although L.K. Blackmore in his book on building the Bentley BR2 rotary engine states that it is a two-man job. It involves heating the steel up as it is wound on to a mandrel to form a spiral.

Valve-Spring Retainers

The valve springs have to act on the stem of the valve in order to close the valve at the correct time and to ensure that the valve movement follows the cam movement exactly.

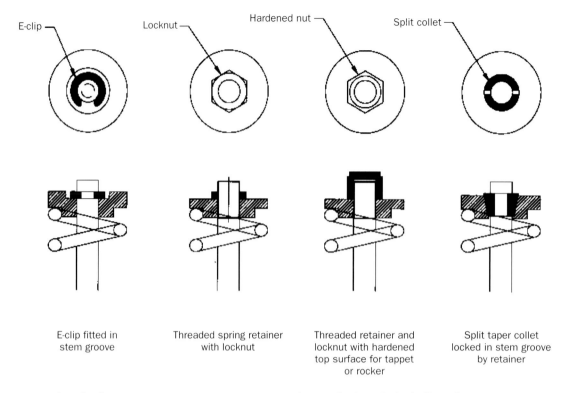

E-clip — Locknut — Hardened nut — Split collet —

| E-clip fitted in stem groove | Threaded spring retainer with locknut | Threaded retainer and locknut with hardened top surface for tappet or rocker | Split taper collet locked in stem groove by retainer |

Examples of valve-spring retaining arrangements. Other methods include drilling the valve stem and using a pin instead of an E-clip and using a hollow nut. The design used will depend on the size of the valve stem, with threaded fittings best for thin stems.

The spring acts on the stem via a retainer, which is located on the end of the stem in some way. It may also provide the valve clearance adjustment, especially in side-valve engines.

There are various types of retainer in use, ranging from a simple nut and locknut on the screwed end of the stem to more complex taper collet arrangements locating in a groove in the stem.

The easiest and most common arrangement is a shallow groove in the stem with a retainer slid over the stem, retained by an E-clip or a small slotted retainer. (When using E-clips, it is important to adhere to the correct fitting tolerances, which will normally be found in the manufacturer's data sheets.) This arrangement is easy to machine and provides good security – you do not want a valve dropping into the engine when it is running! The retainer also keeps the valve spring concentric to the stem and guide.

Whichever method is used, it must be as light as possible because these parts are reciprocating along with the valve and contribute to the oscillating weight.

For all the methods using a groove in the stem, the groove must be a close fit on the fixing to avoid fretting and wear.

The manufacture of the parts is normally a turning job and, for retainers that do not form the tappet, they can be made of anything from HE15 aluminium alloy to steel. Where the retainer also forms the tappet, case-hardened mild steel or hardened silver steel are needed.

TAPPETS

The tappet is the part of the valve gear that transmits the movement of the cam or rocker to the valve stem or pushrod. In some overhead-camshaft engines, the tappet acts directly between the cam and the valve stem. The tappet also absorbs the sideways thrust imposed by the cam and, because the cam acts on the tappet with a sliding motion, the tappet has to be hardened to take the wear.

The face of the tappet that bears on the cam must be of sufficient diameter so that the cam always acts on the face. A tappet that is too small will allow the cam to rub on the edge of the tappet as the tappet is lifted, causing premature wear and early failure.

If space allows, the tappet can be a plain cylinder of suitable diameter but some engines have tappets with a wider face towards the cam (in other words, a top hat shape), which may be circular. In some cases the working face is rectangular, in which case guides must be provided to stop the tappet rotating in use.

Tappets for overhead-camshaft engines with the cam acting directly on the valve stem will often be bored out so that they slide down over the valve stem and spring. This reduces the overall height of the engine and also reduces the weight of the tappet.

Tappets may be made of silver steel or mild steel but must be hardened in both cases. Silver-steel tappets must be hardened right out while mild-steel tappets should be case-hardened. The working surface must be polished before hardening in both cases.

Plain cylindrical tappets may benefit from small oil grooves in the bearing area.

Tappets need to be as light as possible, so it

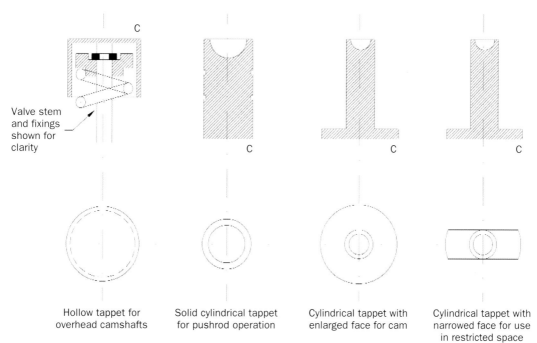

Hollow tappet for overhead camshafts

Solid cylindrical tappet for pushrod operation

Cylindrical tappet with enlarged face for cam

Cylindrical tappet with narrowed face for use in restricted space

Valve stem and fixings shown for clarity

In all cases the cam face is marked "C"

Various tappet designs. The right-hand design must have a guide that prevents it from turning.

pays to bore out the centre to reduce the weight.

ROCKER ARMS

Rocker arms transmit the movement from the tappet or cam to the valve stem and usually incorporate the valve-clearance adjustment. Overhead-camshaft engines will only have rocker arms if the camshaft is set to one side of the valve stem.

The faces of the rocker that act on the valve stem and cam must be hardened to avoid wear. Rockers for pushrod engines will have a locating hole for the top end of the pushrod in the end of the rocker, possibly as part of the valve-clearance adjustment.

The rocker will normally pivot (rock) on a shaft, but other arrangements have been tried. In any case, the pivot is normally equidistant between the working parts of the rocker arm, meaning that the cam movement is exactly replicated at the valve end. If the distances are not equal, the valve timing and lift is affected, and this must be taken into account in the design. The stresses on the rocker may also be higher in this situation.

Valve-Clearance Adjustment

For engines without rocker arms, the valve clearance is often set by means of shims between the tappet and valve stem.

Engines with rocker arms will usually have some form of adjustment built into the rocker arm. This will commonly be a screw adjustment at the pushrod or tappet end of the arm.

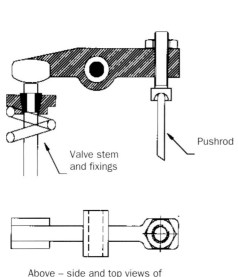

Above – side and top views of standard rocker arm for use with pushrods

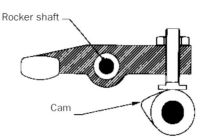

Standard rocker for use with overhead camshaft to side of valve

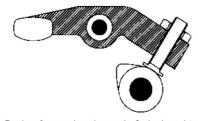

Rocker for overhead camshaft designed to reduce the width of the cylinder head

ABOVE: *Rocker arms and layouts for different camshaft arrangements.*

OPPOSITE: *A half-size working scale model of a twin-cylinder Matchless G45 racing motorcycle engine by Bill Connor. This engine was built without using castings and the valve gear is exactly as the original.*

Some designs incorporate an eccentric rocker shaft, which is turned to lift or lower the pivot point with the adjustment fixed by means of a locknut. One possible disadvantage with this method is that adjusting the rocker will also move it a fraction sideways, although this is not normally significant.

Construction

Rocker arms may be machined or filed from the solid, or fabricated. They are normally made from steel with the working faces case-hardened. If the working faces are made separately, then other materials can be used. One example is an engine with aluminium-alloy rockers with screwed-in hardened-steel inserts that also provide the adjustment.

If rockers are fabricated from steel by silver-soldering or brazing, if the tips made of silver steel, the hardening of the tips can be carried out by quenching the rocker during the soldering process. Alternatively, you may prefer to machine the arms from mild steel and then to case-harden the working faces.

Rockers may be fitted with bearing bushes or even small ball or roller races if desired.

For multi-cylinder engines, it is often sensible to make suitable jigs to hold the rocker arms for machining; this will probably save time and will certainly help to produce consistent results.

One approach is to machine a piece of bar to the outline shape of the rocker and to drill the bearing hole before slicing off the individual rockers with a slitting saw. Surplus metal can then be removed with the rocker clamped in a jig that locates in the bearing hole.

PUSHRODS

The pushrod is no more than a straight rod held between the tappet and the rocker arm to transmit the movement of the tappet to the rocker arm. Pushrods are only used in overhead-valve engines where the camshaft is set low down in the cylinder block; they are also used occasionally in side-valve engines in similar circumstances.

The pushrod is usually held in place by means of the ends being located in the tappet and the rocker arm. Valve-clearance adjustment is sometimes provided using threaded adjusters on one end of the pushrod, but the more normal method is to incorporate the adjust-ment in the tappet or rocker arm.

The easiest way to make pushrods is to use a length of piano wire with the ends rounded and polished to locate in semi-circular dimples in the rocker and tappet. Pushrods made from this material do not need to be very thick as it is a very stiff material and the force is along the length of the rod. As a guide, one 15cc pushrod engine operates with 2mm diameter piano-wire pushrods 110mm long.

Some constructors have used built-up pushrods of aluminium bar or tube with steel inserts in the ends to take the wear, but it is doubtful whether any weight is saved using this method, unless the rods are very long.

12   Cams, Camshafts and Drives

These components provide the correct valve movement in relation to the crankshaft rotation. Together with the valves themselves, they form a critical part of any four-stroke engine.

CAM DESIGN

The purpose of the cam is to open and close the valve at the correct times and as smoothly as possible. The tappet should accelerate and decelerate smoothly throughout the operating cycle, with no shock to the valve gear. The shape of the cam is critical to this function, as is the type of tappet used.

The typical four-stroke engine considered here will most commonly use flat-faced tappets acting directly on the cams. This imposes some constraints on the shape of the cams if the valve gear is to function correctly.

Some engines use a roller tappet, which makes the machining of cams simpler but adds weight to the valve gear. This type of tappet is typical of slow-revving horizontal engines.

The main difference between the shapes of the cams in these two situations is that the cam used with the flat tappet must have curved flanks, whereas the cam for the roller tappet can use straight flanks. The reason for this is that, if

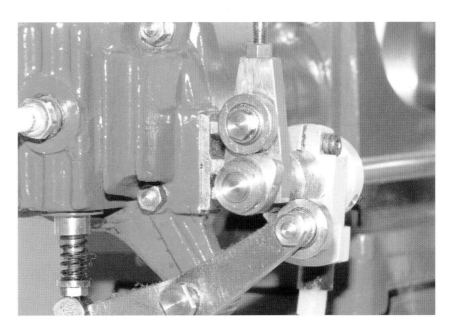

Roller tappets and flat-flanked cams on a gas engine.

flat tappets are used with straight-flanked cams, the lifting action is very sudden and gives a shock load to the valve gear. In such situations, the tappet will tend to 'bounce' away from the cam, with the result that the valve will not open and close at the correct points. Wear is also likely to be a problem in this situation.

If the rocker acts directly on the cam, the rocker face could be curved and used with a straight-flanked cam.

The use of curved and roller cam followers does change the valve timing to some degree when compared with a flat tappet running on the same cam. The opening and closing points and also the path followed by the valve will differ slightly from the cam contour.

The cam contour can be divided into several discrete parts: the base circle, which is the part of the cam in use when the valve is closed; the opening flank, which controls the valve opening; the nose, the part of the cam in use when the valve is being held open around maximum lift; and the closing flank, which controls the cam closing portion of the cycle. In most engines, the opening and closing flanks on each cam will be the same. Some constructors of advanced engines have used different curves in an attempt to gain more performance or economy.

The exhaust cam is often made with a slightly longer opening time than the inlet cam, but this will depend on the engine design and is not essential.

The transition between the different parts of the cam must be smooth, with the different parts of the curve blending into each other.

The drawing of cams is much easier with CAD systems because the majority of such systems have the facility to blend curves into each other. Those using pencil and paper may need several attempts to achieve a good cam shape.

The detailed cam diagram below (see page 37) shows the different parts of the cam for a

Roller must be large enough to avoid point of contact moving too far round the roller circumference

Follower must be wide enough to ensure that point of contact does not move off the follower

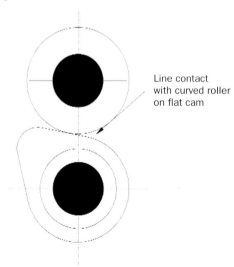

Line contact with curved roller on flat cam

Straight flanked cam with roller follower

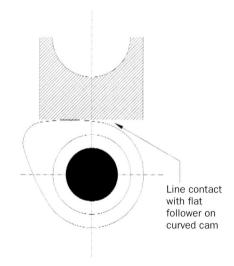

Line contact with flat follower on curved cam

Curved flank cam with flat follower

Action of roller and flat tappets on the correct cams.

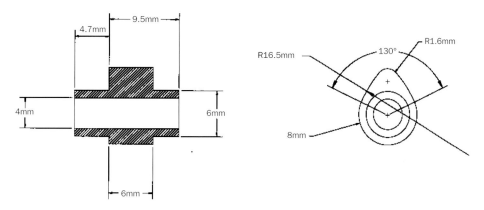

Dimensioned cam diagram from a 15cc four-stroke engine.

single-cylinder engine design. In this cam, the valve is open for 130 degrees of cam rotation – 260 degrees of crankshaft rotation – and has a maximum lift of 2mm. The cam uses a flat tappet and curved flanks with a radius of 16.5mm. The flank curve affects the rate of acceleration of the valve and a good compromise is to use a figure of around twice the base circle (in this case, 8mm).

The nose radius was arrived at by using the CAD program to match the radius to the cam flanks and maximum lift circle diameter of 12mm.

Those designing their own cams will find that changing one of the parameters will affect all the others; for example, altering the opening angle will affect the useable flank and nose radii. If the chosen opening angle and flank radius do not allow the correct lift or nose radius, then making the cam larger will help because the base circle is then larger in relation to the lift, giving more flexibility.

If this is done, the tappet working face may need to be larger to accommodate the larger cam. Such are the trials of an engine designer!

CAMSHAFTS

Once the cam profiles are specified, the overall design and method of manufacture of the camshaft must be considered. These will be constrained by the facilities available and also by the personal preference of the designer.

The basic choice to be made is whether the shaft will be built up or machined in one piece. The cams must have hardened surfaces, and this will greatly affect your decision. For one-piece camshafts, the likelihood of distortion during the heat treatment process is high; this means that those without some form of heat-treatment furnace to provide an even, steady heat may be best advised to use a built-up camshaft.

With built-up camshafts, the cams are manufactured individually and then fixed to the shaft (which is typically of silver steel) using Loctite or similar. This method is by far the easiest to use and there are never any problems with cams shifting on the shaft. Another advantage of built-up shafts is that intermediate ball or roller bearings can be incorporated easily, provided care is taken to avoid getting Loctite into the bearings during assembly.

Cams may also be made with provision for grub screws to fix them to the shaft. The screws should be proper socket screws with hard tips.

The other factor that may affect your choice is access to cam-grinding facilities. If such facilities are available, the hardening process can be carried out before final grinding, meaning that any distortion can be removed during the grinding process. For one-piece

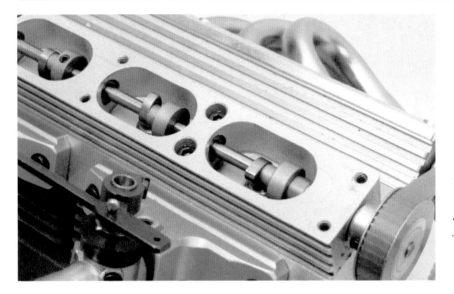

Built-up camshaft in a four-cylinder engine showing the grub-screw fixings for the cams and the intermediate bearings, which are needle rollers.

shafts, silver steel or some other grade of steel that can be hardened is used. For built-up camshafts, case-hardened mild-steel cams fixed to a plain unhardened silver-steel shaft is the common combination of materials.

CAMSHAFT DRIVES

The camshaft drive transmits the drive from the crankshaft and reduces the relative speed by a factor of 2:1 in the four-stroke engine.

The choice of drive method will depend on the engine design and the distance between the two shafts. There are three options available to the constructor: gears (including skew gears), chain drive and toothed belt drive.

Drive sprockets and gears can be held on to the crank and camshafts using tapers or taper collets, preferably with some form of small key as a positive location and for ease of reassembly of the engine.

Geared Camshaft Drives

Spur-geared drives are very commonly used on both side-valve engines and pushrod overhead-valve engines. This is because the distance between the camshaft and crankshaft is typically small, meaning that two or three gears

can easily bridge the gap without the individual gears being very large. Remember that if more than two gears are used in this way, the 2:1 ratio is only dependent on the relative number of teeth on the first and last gears.

A three-gear arrangement has the advantage that the middle gear can be mounted on an adjustable shaft so that the correct meshing of the gears is easy to achieve. One advantage of using gear drives is that the gears can often be cut by the home constructor, thus saving money and also meaning that the choice of gears is not dependent on commercial availability.

Although gear drives are the norm for side-valve and pushrod engines, the Westbury four-cylinder overhead-camshaft *Sealion* design uses a spur-gear drive with two intermediate gears, giving four in all.

Skew-gear drives are most commonly used on horizontal engines but Edgar Westbury originally specified such a drive on his *Kiwi* design, although the Mark II version of this engine uses spur gears. The problem with skew gears is usually that of obtaining the correct mesh with two shafts positioned at right-angles to each other. Such gears are also difficult to cut for the home user and can be hard to find and expensive to purchase.

Spur-gear camshaft drive for a single-cylinder pushrod engine.

One point to make with all geared drives is that the commercial gears for the drive may need to be modified, usually by altering the bore size. If this is the case, a split collet should be used to hold the gear for machining. It is essential that the gears run truly on the shaft, to avoid noise and wear.

Gear drives need to be enclosed and provided with some form of lubrication. This is often done by means of a vent into the crankcase, which allows oil mist to reach the gears.

Chain Drives for Camshafts

In the past the chain-drive camshaft was by far the most common found in full-size engines. It has now largely been superseded by the toothed belt drive.

In model sizes, the chain drive can be a good method to use for overhead-camshaft engines, providing that small-size chain can be sourced. The sprockets can be cut by the constructor and should have the teeth hardened, although, if chain can be sourced, suitable sprockets will no doubt also be available.

Chain-drive components for camshafts.

One issue that may arise is that of the increments available for chain lengths. If the chain pitch is 10mm, that will be the minimum increment of length. This is something to be considered at the design stage, since a small variation in the distance between the pulleys will make a significant increase in the amount of slack in the chain to be taken up by the tensioner. It may not be possible to achieve this if the increment is too large.

Chain drives need a tensioner to take up the slack in the chain and this should be on the back side of the drive. This is the side that does not take the pull of the load, so, looking face on to an engine that runs anticlockwise, the tensioner must be on the right-hand side.

Like gear drives, chain drives must be adequately lubricated.

Toothed Belt Drives

Toothed belts are the modern equivalent of the chain drive and offer several advantages: they are quiet, require no lubrication and are cheap in small sizes.

The toothed belt is moulded from either neoprene rubber or polyurethane with steel or glass-fibre tension members moulded in. For miniature engine use, either is suitable; polyurethane with steel inserts seems to work particularly well.

Toothed belt and pulleys, widely available commercially.

The belts and suitable pulleys are easily obtainable down to small sizes, although the pulleys may be home-made if desired. If this is done, it is important that the correct tooth form is used, to avoid wear on the belt with possible failure.

Toothed belts can be obtained down to pitches of 2.5mm but the available increments of belt length may be greater than this, so it is important to check the available belt lengths at the design stage.

In miniature engines, a correctly fitted toothed belt should last the life of the engine, although, because of the low cost, if there is any doubt about the state of the belt it is worth fitting a replacement.

Care should be taken to ensure that the belt has adequate 'wrap' round the pulleys, particularly the smaller crankshaft pulley. Plain pulley tensioners can be fitted to run on the outside, normally flat, surface of the belt.

Belts can be obtained with teeth on both sides, which may be used to drive water and oil pumps. If this is done, however, it is important to take care that the belt is not overloaded. Usually, it is better to use a separate belt or other means – particularly where you need to drive an oil pump.

Toothed belts should be tensioned so that there is a minimum of sideways movement in the belt between the crankshaft and camshaft pulleys, but they should not be over-tightened. As an example, a toothed belt drive might have about 5mm of sideways movement in a length of about 100mm.

Unlike gears and chains, toothed belts do not need any lubrication and do not need to be fully enclosed, other than for safety reasons.

One point to make is that the toothed belt does not need to be large for miniature engines; two examples are a 40cc four-cylinder twin overhead-camshaft engine and a 15cc single, both of which use a belt of 2.5mm pitch and 6mm wide for the camshaft drives. A glance at the belt manufacturer's specifications will show that the load that can be transmitted by this size of belt is more than adequate in both cases. Many engines are spoilt by the fitting of oversize belts.

In all cases, once an engine has been set up and timed correctly, the gears and pulleys should be marked with the piston at top dead centre (use the first cylinder in multi-cylinder engines), so that the valve timing can be easily reset if the engine is disassembled for any reason.

MAKING CAMS AND CAMSHAFTS

There are two main methods of making camshafts: they may be built up or in one piece.

Within these two methods there are several variations, for example, using a purpose-built cam-cutting and grinding fixture; using a jig in the lathe and finishing by filing as per Edgar Westbury; or using a set of computer-generated cam lift figures to cut the cams in the milling machine using a rotary table. All three methods can be used for built-up or one-piece camshafts, although the Westbury jig can be quite complex to set up with multi-cylinder engines for one-piece shafts and care is needed if mistakes are to be avoided.

Cam-Cutting and Grinding Machines

Those wishing to use cam-cutting and grinding machines will come across a significant problem: they will have to construct the machine themselves. Several designs have been published in the model engineering press over the years.

The operating principle of such machines is that a large copy of the cam profile (typically cut from thin sheet metal) is driven from the rotating camshaft blank, and this follower is connected through a system of levers to the cutter or grinding wheel, causing it to move in line with the cam profile.

Some machines move the cam blank rather than the grinding wheel or cutter, but the principle is the same.

The cam template is cut to the correct profile, but several times larger depending on the set-up of the machine. This means that it

can be accurately marked out and cut by hand since the reduction built into the machine will reduce the effect of any minor imperfections in the profile. Additionally, all the cams will be identical because they are cut from the same template.

If the machine is designed to use a milling cutter for the initial profiling, it needs to be very rigid, implying heavy construction throughout. Grinding loads are lighter, so the machine can be lighter, although producing cams by grinding only will take longer.

In order to cut all the cams on a camshaft, some means of indexing the camshaft blank relative to the template between cams must also be provided. It is often easiest to provide indexing at the template rather than at the cam blank.

For those who wish to construct one-piece hardened camshafts, the cam-grinding machine is probably an essential because the cam profiles can then be cut with a grinding allowance built in. The shaft is then hardened and returned to the machine for final grinding and polishing.

In use, a blank shaft is turned up with the cams as circles of the maximum lift radius (or a fraction greater) where each cam will be, and these are then cut or ground to the correct profile in the machine.

Cutting in the Lathe

The method given here of cutting cams in the lathe was devised by Edgar Westbury many years ago, before most model engineers had access to good-quality milling machines. It involves using a jig to rotate the cam blank at a suitable radius to turn the cam flanks. The jig also has indexing facilities so that the correct relationship between both the flanks and the different cams on the shaft can be set.

The disadvantage of this method is that the base circle and nose radius are finished off by filing, but, provided care is taken, good cams can be achieved.

One-piece camshafts or individual cams may be produced by using this method, although the jig will be different in each case, and a new

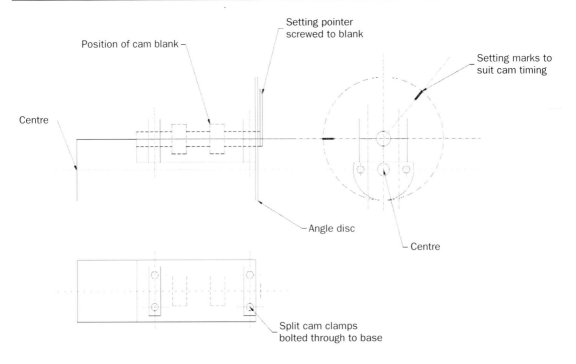

Edgar Westbury-type cam-turning jig for use in the lathe.

jig will be needed for each engine.

This method may be used to cut individual cams for both four-cylinder and single-cylinder engines. In the case of the four-cylinder engine, cut the exhaust and inlet cams as a block and file them to final shape before parting them off to fit to the shaft. It is easier to maintain the correct shape when filing the wider block than it is for individual cams. The cams are also more likely to have a consistent profile by doing this. The cams for the single-cylinder engine may be made in the same way as long as it is designed with identical profiles for the exhaust and inlet cams.

The procedure is as follows:

1. Set the cam blank in the jig with the pointer at the first valve-opening mark.
2. Bring the tool up to the blank surface with the cam at its nearest point to the tool and zero the cross slide index.
3. With the lathe running at slow speed, take light cuts until the tool has been advanced to give the correct base circle diameter. Zero the cross slide index at this point.
4. Rotate the blank and pointer to the cam-closing mark and take light cuts down to the same setting.
5. Mark the small flat (the nose) between the two flanks to avoid confusion later.
6. Now index the blank to take a series of cuts at small increments round the base circle to remove as much metal as possible. These cuts go to the same depth as those for the flanks.
7. Move on to the next cam and repeat the process.
8. When all the cams are cut, remove the blank and finish the profiles by filing, leaving a fine polished finish. Use a template to check the nose contour if possible and ensure that the flank curve is not destroyed during this process.

Milling

With the use of a computer program to calculate the required data, milling is now one of the best ways of producing cams. Such programs have been published in the model engineering press in the past, or go to **www.modelenginenews.com** for an online example.

The program produces a set of 'offsets' for each degree (typically) of cam rotation, which is effectively the lift of the cam at that point. This means that an end mill can be used to take cuts at each setting starting from the base circle diameter.

It may be thought that this method would produce a series of ridges across the face of the cam; however, these ridges are very hard to see and are easily polished off before the cams are hardened.

The advantage of the method is that all parts of the cam are accurately cut and the results are extremely consistent, which is important for multi-cylinder engines.

The requirements for milling cams are a vertical milling machine with a rotary table or a dividing head and an accurate method of indexing the cutter by very small amounts (as low as 0.01mm) between each cut. A digital read-out fitted to the mill is ideal. If you do not have this, using a dial gauge mounted under the head is the best method. It is not normally sensible to rely on the machine indexes on the average model engineer's mill because the increments between cuts are so small.

The procedure is as follows:

1. A cam (or camshaft) blank is machined and set up in the rotary table to run absolutely truly in the horizontal axis. It is advisable to support the end of the shaft with a centre.
2. The rotary table is set to the first cut figure on the list of offsets. This is often not zero; it is usually set a couple of degrees before the valve-opening point.

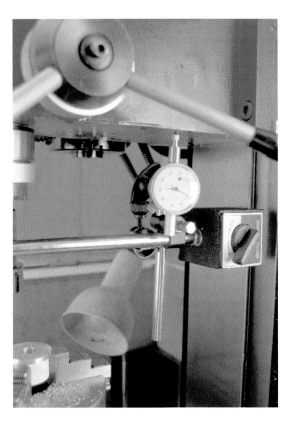

Using a dial gauge to index the milling head for very small cuts in the absence of digital read-outs.

Milling cams using the rotary table on the vertical milling machine.

3. The cutter is brought down to the blank and the read-out or dial gauge is zeroed. The blank is then machined down to the base circle diameter using the index, and the index is zeroed at this point.

4. From this point on, nothing must be changed other than setting the offset figure and the rotary table for each cut.

5. The rotary table is advanced by one degree, the required offset is set and another cut is taken.

6. This is repeated for the complete cam. Note that, once the valve-closing point is reached, the offset will have returned to zero (in other words, the base circle) and the remaining cuts back to the starting point will be at that setting.

7. The process is repeated for the other cams on the shaft, with the correct starting point being set on the rotary table for each cam.

8. The cams are then polished using fine files (diamond files are good for this) before being hardened and finished.

CNC Machining

Those with access to CNC milling facilities can obviously produce cams using this method. It is advisable to use the finest increments available for the machine in this case.

There are companies who will produce CNC components from standard CAD drawings and these could be investigated by those with the capacity to produce CAD drawings to one of the recognized standards.

Hardening Cams and Camshafts

Cams and tappets need to have a very hard surface on the working areas and this is accomplished by case-hardening, or the use of heat treatment in conjunction with a suitable grade of steel.

For those wishing to harden silver-steel (or

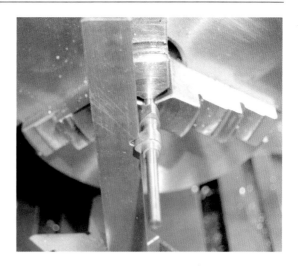

Assembling a cam shaft with Loctite adhesive, using the rotary table in conjunction with a simple jig in the milling machine to set the correct angles.

other suitable grade) camshafts, it is desirable to use a furnace in which the shaft can be brought up evenly to the required temperature before being quenched. Even then, distortion may take place during the process.

The best method is to harden the cams separately and fix them to the shaft using Loctite. When using this approach, the cams should be hardened by covering them completely with case-hardening compound in a suitable container and then heating the whole lot to the required temperature before quenching the cams in water. The cam surface should be a light grey colour and must be glass hard.

The process can be repeated to build up a greater depth of hardening if desired.

Once hard, the cams can be polished with fine emery paper before degreasing and fixing to the shaft. Using a suitable jig will ensure the correct relationship between each cam.

13   Lubrication

All internal combustion engines need some form of lubrication in order to function properly. In addition to providing lubrication, the oil acts as a coolant and, with pumped lubrication systems, it removes heat from the inner parts of the engine.

In two-stroke engines, lubrication is achieved by mixing oil with the fuel, the oil being deposited on the working parts of the engine as the fuel mixture passes through the crankcase.

In four-stroke engines, the situation is somewhat different because there are more parts to be lubricated, and the fuel mixture does not normally pass through the crankcase on its way into the cylinder. The common exception to this is in rotary and radial engines, where the mixture is taken through the crankcase and can therefore be used to carry oil through in the same way as for two-strokes.

Slow-revving horizontal engines typically use oil or grease cups on the main bearings and big end in conjunction with an oil cup, to provide cylinder lubrication. The cups are charged with oil or grease as needed.

For a typical four-stroke engine, the most common lubrication options are splash, pumps of various types, crankcase suction and oil in the fuel. Your choice will depend on the engine and the type of bearings used.

SPLASH LUBRICATION

Splash lubrication was commonly used on early motor engines and several of the Westbury designs incorporate this method.

The basic principle requires the sump to be filled with oil to the level of the lowest point of big ends, so that big ends dip into the oil as they rotate. As well as lubricating the big ends via a small oil hole in the bottom of the bearing, this action causes the oil to be distributed around the inside of the engine, thus lubricating the other parts.

Lubrication of overhead camshafts is often by means of a breather pipe up into the cam box, which allows oil mist to reach those parts.

Frequently, a semicircular trough is fitted under the big ends to provide better control of the oil pick-up. In some cases, troughs are fitted above each main bearing to catch oil running down and to feed it into those bearings.

Splash lubrication is an easy method to use and will provide perfectly adequate lubrication in most medium-performance engines. It is not suited, however, to engines that are subject to violent movement, as in model aircraft, or for engines in model boats, which may be mounted at an angle to line up with the propeller shaft.

For best results, one of the thinner engine oils should be used so that it will reach all the parts of the engine.

PUMPED LUBRICATION

Full-size engines invariably use an oil pump to distribute oil to the working parts of the engine and many miniature engines have been constructed with such systems. The pumps used are of two main types: gear and plunger.

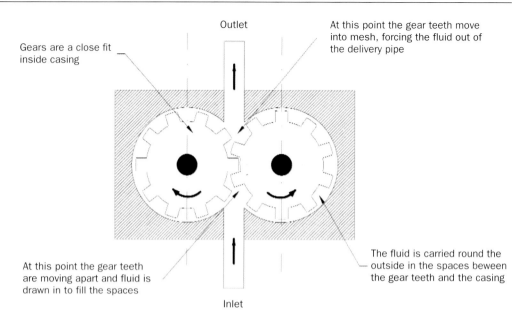

Outlet

Gears are a close fit inside casing

At this point the gear teeth move into mesh, forcing the fluid out of the delivery pipe

At this point the gear teeth are moving apart and fluid is drawn in to fill the spaces

The fluid is carried round the outside in the spaces beween the gear teeth and the casing

Inlet

Principle of gear pump operation.

The latter may be either the reciprocating or oscillating type.

A more reliable oil feed is often achieved if the pump is mounted below the oil level in the sump so that gravity feeds the oil to the inlet port.

Gear Pumps

Gear pumps are commonly used on full-size engines and even in small sizes are capable of pumping large quantities of oil at high pressures. For this reason they are almost always fitted with a pressure relief valve to allow excess oil to be diverted back to the sump. Properly made gear pumps will provide a very reliable feed and are generally better at handling air in the system.

Gear pumps must be made accurately and with a close fit between the casing and the gears. The gears should be set to mesh as tightly as possible, while maintaining free rotation. For this reason the chamber for the gears in the casing is best bored rather than drilled and milled.

The action of the pump is as follows: the oil enters the pump at one side of the gears on the centre line; it is carried round the outside between the gears and the casing (filling the gap between the gear teeth), until it reaches the point where the gears mesh; the oil is then forced out by the meshing action and leaves by the only exit, the outlet pipe.

In model engines, the pump is often placed outside the engine and can be driven off the camshaft or by an independent drive from the crankshaft. Pumps can be mounted inside the sump if this is practical, but the usual problem is in obtaining the drive.

It is a good idea to put a fine mesh filter over the end of the intake pipe to avoid any foreign matter finding its way into the pump and hence to the bearings.

The oil feed pipe is usually fed into one of the main bearings and oil reaches the other bearings via oil-ways drilled in the crankshaft. In multi-cylinder engines, drilling these oil-ways can be quite tricky, requiring long thin drills and great care, to avoid drill breakage when the drill cuts through into other oil passages. You will need to give some thought as to the best way to drill the holes, the ends of which will need to be plugged once all is drilled. This may be done with a threaded plug screwed in with a smear of Loctite. The surface of the plug can be cleaned up after insertion; it is important to make sure that all swarf is removed from the holes before this is done.

Pressure relief valves can be made in the same way as those used on steam locomotives, using a spring-loaded ball or wing valve set to the correct pressure. They can be mounted in the main oil feed pipe wherever this is convenient, with the overflow being returned to the sump.

Small oil pipes can be taken up to overhead camshafts, or oil mist can be relied on, as with splash lubrication.

If oil pipes are used, they will need a needle valve in order to be able to adjust the flow to these areas while maintaining the correct flow to the other bearings.

Reciprocating Pumps

Reciprocating pumps can be of the oscillating type or various cam or eccentric driven plunger types. Both types work well but some designs of reciprocating types are quite complex, because

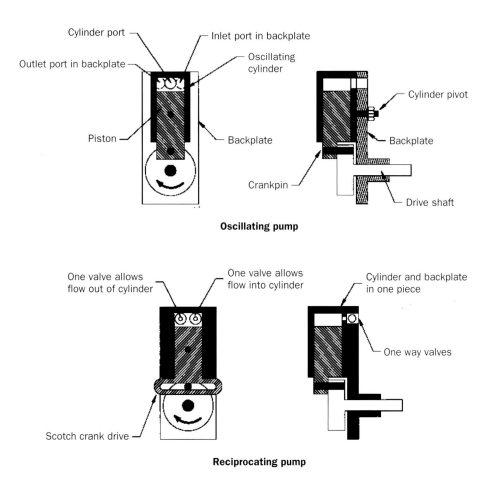

Oscillating pump

Reciprocating pump

Reciprocating and oscillating oil pumps. With both types, the back plate may be part of the engine structure. Both types can be driven from the camshaft or crankshaft, often using skew gears.

the pump ram is shaped and oscillates in use to control the inlet and outlet. Straight plunger pumps will need non return valves and, if they are above the level of the oil, they may not pick up as well as the other types if air gets into the system.

The volume of oil pumped by this type of pump is generally less than that from gear pumps and some constructors do not fit relief valves in this case. This is especially true with the oscillating type, in which the spring holding the port faces closed will act as a relief valve. (The oscillating pump does not need a heavy or long spring to hold the port faces together, in fact, a spring washer will suffice and this will keep the size of the pump down.)

These pumps can be mounted inside or outside the engine but oscillating pumps are best mounted inside so that any oil escaping round the port face stays in the engine.

Drives for these pumps are often via skew gears off the crankshaft, which suits the layout of such pumps.

CRANKCASE SUCTION LUBRICATION

Some early engine designs used the variations in crankcase pressure caused by the piston movement to pump (or, more correctly, suck) oil into the bearings. This method of lubrication is particularly applicable to single-cylinder or horizontally opposed twin-cylinder engines, in which the pumping action is substantial.

In order to use this method, the crankcase must be sealed, as in two-stroke engines. It must also be fitted with a one-way valve, which will allow air to escape as the piston descends, but which prevents air entering as the piston rises, thus causing a drop in pressure in the crankcase. It is this drop in pressure that is used to suck oil from an external tank into the engine, usually via oil-ways in the crankshaft and through the big-end bearing, from which it finds its way to the rest of the engine.

One problem with this method of

lubrication is that it is a total-loss system because the oil either stays in the crankcase or escapes through the one-way valve.

The entry through the main bearing should be through a hole in the crankshaft that lines up with the oil feed as the piston approaches the bottom of the stroke. This will allow more than enough oil into the engine.

This method of lubrication will work with the engine at any attitude as long as the feed pipe in the oil tank remains below oil level.

OIL IN THE FUEL LUBRICATION

This method of providing lubrication in four-stroke engines is particularly applicable to glow-ignition engines, for two reasons: first, there is no spark plug to oil up and, second, commercial glow fuel will almost certainly contain oil.

The crankcase one-way valve for crankcase-suction lubrication on a single-cylinder engine. The actual valve is a small free-floating paxolin disc.

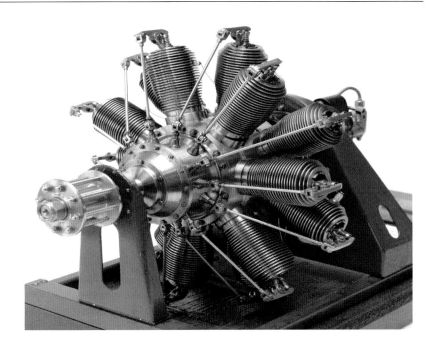

The unusual double row Gnome rotary engine by Les Chenery. Lubricated with oil in the fuel, a total loss system.

Some engines take the mixture through the crankcase and into the cylinder, which means that the crankcase must be sealed, but many others rely on the blow by of oil past the piston to lubricate the engine.

Because the fuel contains oil, this also helps to seal the piston and maintain compression, even though the clearance in such engines may be greater than in the equivalent spark-ignition engine. Some modern commercial glow engines have a slightly tapered bore, which means that the clearance at the bottom of the stroke is greater, thus allowing oil to pass by to lubricate the bottom end of the engine.

The crankcase is often vented at the bottom to allow the excess oil and the corrosive by-products of combustion to escape.

This lubrication method particularly suits engines with ball or needle-roller bearings for the main and big ends because these need only oil mist for proper lubrication, although many plain bearing engines also run successfully.

TYPE OF OIL TO USE

The type of oil to use in four-stroke engines will depend on the type of ignition and also the type of fuel used.

For spark-ignition engines with splash or pumped lubrication, any low-viscosity motor oil, either synthetic or normal, will be fine.

For engines with lubrication using oil in the fuel, commercial two-stroke mixing oil can be used, with a percentage of oil to fuel of around ten per cent.

Glow-plug engines use methanol-based fuels and in this case the best oil to use is a synthetic or castor-based oil, which will mix with the methanol. Special oils for use in glow fuel can be obtained from model shops dealing with commercial glow-plug engines.

On no account should synthetic and ordinary types of oil be mixed in an engine. This can lead to the formation of sludge, which can block oil-ways and cause serious problems.

14   Engine Balance

All engines need to be balanced in order to avoid vibration and to develop maximum power. This is particularly true of single-cylinder engines, and also applies to multi-cylinder engines, although to a lesser extent.

With single-cylinder engines, the problem is obvious because of the weight of a piston and small end moving up and down in a cylinder, which causes vibration. This can be balanced out to some extent but there will always be some vibration inherent in a single-cylinder design.

With multi-cylinder engines, although the pistons apparently balance each other out, there will still be some torsional vibrations set up, due to the fact that the pistons are not directly opposite each other in most engines.

The engines with the best inherent balance are flat-twin or flat-four engines with opposing connecting rods on the same big end. It is noticeable that a flat-four will generally achieve higher rpm figures than an equivalent four-cylinder in-line engine.

THE PROBLEM AND THE BALANCE COMPROMISE

The balance problem is best illustrated with the worst example: a single-cylinder engine (*see* diagram, right).

The diagram shows how, with the piston at top dead centre, there is a force tending to push the engine vertically upwards as the piston direction reverses. The same is true of the situation at bottom dead centre, although here the force is downwards. At each end of the stroke, therefore, there is a force acting on the engine, hence the vibration in single-cylinder engines.

Obviously, these forces could be balanced out by having a counterweight on the opposite side of the crank web that will produce a force equal and opposite to that due to the piston (plus the gudgeon pin and connecting rod) at the designed rpm. This is fine, but near the half-way point in the stroke, when the crankpin is at right-angles to the cylinder centre line, this counterweight will produce the same force as before but it will not be equalled by an equivalent force. This is due to the piston and connecting rod, which are constrained in the cylinder and are, in fact, just moving and not creating any appreciable forces at this point.

Unless extra balance shafts create counter-balancing forces at this point, the balance of the engine becomes a compromise. There is a need to cancel out as many of the forces due to the

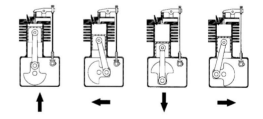

The out-of-balance forces on a single-cylinder engine with a crankshaft balance weight equal to the total of the rotating weight and half of the reciprocating weight.

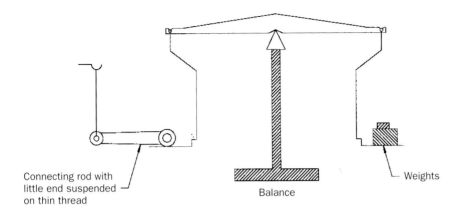

Weighing a connecting-rod big end using a small balance. The connecting rod is suspended by the small end to weigh the big end, and vice versa.

piston and other parts changing direction at the top and bottom of the stroke as possible, but without introducing sideways forces, which may be as bad as, or worse than, the original problem.

Balance shafts have been used to improve balance in full-size engines, but not in miniature engines. The technique involves having rotating counterweights on a separate shaft driven from the crankshaft, which cancel out the sideways forces without adding to the forces at the top and bottom of the stroke.

Most engines rely instead on a compromise balance to keep the vibration to the minimum and this has led to a widely accepted balance process that takes account of the reciprocating and rotating weights in the engine to arrive at the best option.

BALANCING TECHNIQUES

The basic rule used in balancing a single-cylinder engine is that 'the crankshaft counterweights should balance out all the rotating weights and half the reciprocating weights in the engine'. This gives the best overall balance and will generally result in a smooth-running engine.

First, it is necessary to sort out the parts that

fall into each category. This can be confusing because parts such as the connecting rod fall into both categories. In fact, the big end (including any bolts and bearings) can be seen as a rotating weight while the small end is a reciprocating weight. This means that the rotating weight will comprise the big end and the crankpin (although it must be remembered that the crankpin weight is only that part outside the crank web; the part inside the web is balanced by the metal in the equivalent position on the opposite side of the web). It is usually easier to turn a small pin that has the same dimensions as the protruding part of the crankpin and to weigh this to find the weight of this component.

The reciprocating weight consists of the small end, the piston with rings and the gudgeon pin.

In practice, the parts are weighed and the calculated weight is added to the crank web opposite the crankpin. Weighing the parts such as the piston and gudgeon pin is obviously easy but for the connecting rod each end has to be weighed separately. This is best achieved by using a small balance and suspending the big end using thin thread while weighing the small end, and then reversing the process to weigh the big end.

It is not necessary to get absolute weights in grams or ounces – this is an issue of relative weights – so a simple balance can be made and small washers or other identical items can be used as weights.

The existing counter-balance weight can be measured by setting up the crankshaft on horizontal knife edges and suspending weight from the crankpin until it just balances the balance weight. Remember to add the crankpin weight to the final figure in this case.

Once all the parts are weighed, the balance weight necessary can be calculated and compared with the weight found using the weighing process.

Extra weight can be added in the form of steel weights formed as arcs to match the outline of the crank web, but care must be taken to ensure that they do not foul the connecting rod or other parts. If the balance weight is too large, small holes can be drilled to lighten it.

Some crankshafts have separate balance weights bolted to the crankshaft rather than machined webs. In this case, the weight is easily changed by making a new part. Such weights are best bolted on using high-tensile socket screws.

Remember that for engines with a crankshaft that has two webs and shafts at each end, both webs must be taken into account. This is true even if the shaft is in two parts, with the second part running in its own bearings but driven from the crankpin, as when the camshaft drive is taken from the rear of the engine.

FLYWHEELS AND BALANCING

The use of a flywheel will also promote smooth running of an engine due to the extra weight helping to dampen down the vibrations. The problem is that, again, a compromise must be made between using a heavy flywheel, which will dampen down the vibrations and will also cause the engine to accelerate more slowly when the throttle is opened, and the lighter flywheel, which will alleviate that problem but will not provide such a good damping effect.

Flywheels are also not ususally suitable for aircraft engines where weight is normally at a premium.

One other point to be aware of is that the flywheel is generally on the outside of the engine and is therefore not close to the source of the out-of-balance forces, which reduces the effect slightly.

BALANCING MULTI-CYLINDER ENGINES

Multi-cylinder engines are generally better balanced than single-cylinder engines due to the fact that the crankpins are normally arranged opposite each other, thus cancelling out most of the vibrations. It may still be sensible to add some balance weight because the opposing forces will act at different points along the crankshaft, which will mean that there are still some out-of-balance forces being generated. The stresses on the crankshaft will also be reduced if the balance is better.

In general, four-cylinder in-line engines will have the two centre cylinders on one crankpin axis, with the outer cylinders on the opposing axis. The out-of-balance forces in this case will be tending to bend the crankshaft.

Six-cylinder engines have the crankpins spaced at 120-degree intervals, again in opposing pairs, which will provide good balance.

In-line twin-cylinder engines can have opposing crankpins, which obviously provides better balance but in some four-stroke engines the crankpins are in line in order to provide a more even firing pattern. In this case each cylinder should be balanced as for a single-cylinder engine. This is yet another compromise in engine design.

Flat-twin and flat-four engines are well balanced and should not need any special treatment in this respect, although those seeking the perfect engine may well provide some balancing for each cylinder.

15 Carburettors

The carburettor is a critical part of any engine and is often the most difficult item to set up correctly. For this reason, many engine builders today use commercial carburettors from the wide range of small engines on the market. This is often a good idea for a first engine because it will largely eliminate fuel mixture problems. Once the engine is set up in this way, there is nothing to stop the builder experimenting with other carburettor designs – with the knowledge that the engine does at least run. Some early engine designs did not include a carburettor on the engine drawings; perhaps this was because of the difficulties experienced with such things!

The objective of the carburettor is to provide the correct fuel/air mixture for the engine at all power and rpm levels. This means that it has to atomize the petrol or methanol into fine droplets and then mix those with air in the correct proportions so that the mixture will burn efficiently when compressed in the cylinder. This ratio is theoretically 14.7:1 air to fuel by weight for complete combustion, but in practice slightly more fuel is needed for maximum performance because the fuel will not be perfectly mixed with the air in the cylinder.

The carburettor also incorporates a throttle for controlling the speed of the engine, usually by restricting the air flow into the engine.

Most miniature engines incorporate an adjustable fuel valve, so that the proportions of fuel and air may be varied for starting, and also to cater for varying atmospheric conditions. This has the added benefit of enabling the mixture strength to be adjusted without having to replace the fuel jets, as with fixed-jet carburettors.

The most common form of valve is the needle valve in which a tapered needle is moved

A simple spray-bar carburettor with integral fuel tank.

111

Two barrel carburettors: (left) a commercial item with separate slow-running mixture control, and (right) a basic version with simple air-bleed idle adjustment.

into or out of a small jet by means of a screw, thus varying the effective size of the jet and, hence, the amount of fuel it can pass.

One of the problems inherent in miniature engine carburation is that of applying compensation for the varying speed of an engine as the load varies. This is not due to variation of throttle opening but because of the variation in air flow caused by the change in engine speed that occurs when the load on the engine varies. The problem is more critical with increasing load because, as the engine speed drops due to the increasing load, the air flow through the carburettor reduces. This reduces the amount of fuel/air mixture entering the engine; there is a consequent reduction in power, and this is the last thing you want when the load is increasing. Various means have been tried in the past to solve this problem and full-size engines have used variable jets controlled by the manifold vacuum. One example of this is the well-known SU carburettor, which has the needle mounted in a piston that moves up and down according to the manifold pressure.

TYPES OF CARBURETTOR

Over the years, many types of carburettor have been tried on miniature engines, ranging from the simple plain tube with a spray bar to quite complex mixing carburettors with compensation and throttle arrangements.

Simple Spray-Bar Carburettor

For engines designed to run at constant speed and with a fairly constant load, the simple spray-bar carburettor is ideal.

In this type of carburettor, the only way of controlling the speed of the engine is by varying the fuel mixture. More fuel than the ideal will slow the engine but may also cause unreliable running if the mixture is made too rich and, if any oil is used in the fuel (such as for two-strokes), may oil up spark plugs. In glow engines, richening the mixture may well cause the glow plug to cool off too much and the engine to stop.

Barrel Carburettor

The barrel carburettor is probably the most commonly used on small engines, both by home constructors and commercial manufacturers. It has the merits of being simple to make, yet in basic form providing a good degree of throttle control.

The throttle barrel is a rotating cylinder that has the inlet passage running through it at right-angles to the centre line. If the barrel is rotated, the edges of the bore progressively shut off the flow of air through the carburettor. The spray bar is positioned in the centre of the barrel, with the mixture controlled by a needle valve.

Barrel carburettors are often fitted with a separate slow-running mixture control and air-bleed controls for fine adjustment of the mixture. These are capable of throttling down to slow speeds reliably and, just as important, they will pick up speed smoothly once the throttle is opened.

The barrel carburettor is suitable for almost any engine and commercial varieties are available from many engine suppliers and manufacturers.

The barrel carburettor has largely replaced the butterfly valve(s) used in some early engines mainly because it is easier to make in small sizes (although an equivalent double butterfly carburettor can be easier to set up).

Compensating Carburettors

This type incorporates a wide variety of designs, ranging from the early Westbury compensating types through to the modern small carburettors often fitted to small chainsaw and garden-tool motors. The modern carburettors have been used on a variety of larger home-built engines, including 150cc V8s, but are not available in small sizes.

The Edgar Westbury carburettors used a variety of compensation devices including a floating piston, which, when the air flow

through the carburettor dropped, restricted the air passage and ensured that speed of flow over the jet was maintained. Another option was that fitted to his *Seal* design, which had the jet communicating with a small passage that entered the centre of the throttle barrel. This was a method of ensuring that the mixture was correctly controlled because it let in extra air via this passage in relation to the air flow through the carburettor. In this design the barrel is made with a slightly larger aperture at the engine end, which increases the air flow through the auxiliary passage as the throttle is closed, thus altering the mixture. These carburettors require proper setting up to give good results, and this is not always easy, but they do provide a degree of compensation for varying load.

The modern commercial compensating carburettors are quite complex. Many incorporate a fuel pump driven off the inlet vacuum pulses and have varying degrees of compensation built in. A typical example of such a carburettor is the Walbro version.

Another method of mechanical compensation is to have the throttle barrel threaded or

A commercial Walbro carburettor suitable for larger engines. This has a built-in fuel pump driven by the inlet manifold vacuum.

grooved so that, as the throttle is opened, the needle is lifted to provide a richer mixture. This allows the engine to pick up speed cleanly because extra fuel is provided to compensate for the increased air flow through the carburettor. The commercial carburettors by Super Tigre, among others, adopt this principle, with the throttle barrel being fitted with an angled slot so that the needle moves away from the jet as the throttle is opened.

This type of carburettor may also be provided with a separate slow-running mixture control, which comes into action only when the throttle is closed.

All of these carburettors may need experimentation to give the best results.

Butterfly Carburettor

The butterfly carburettor is very like those fitted to full-size car engines before the advent of fuel injection and uses one or two flaps (or butterflies) in the intake tract to control the air flow. Compared to the barrel type, they are slightly more difficult to make but have some attributes that may be of benefit.

First, if a single butterfly is used (and the fuel is introduced into the edge of this), as the throttle closes, the air flow between the edge of the flap and the inlet pipe increases in speed, thus maintaining the suction at the jet. This provides a degree of compensation, giving a good mixture at all throttle settings.

Another option with butterfly throttles is to use two coupled flaps, one on the inlet side and one on the engine side of the spray bar. This operates like a barrel throttle, closing off the inlet passage on both sides of the jet. The advantage is that, if some adjustment mechanism is fitted to the linkage, the relative size of the openings can be adjusted. If a richer mixture is needed for low throttle settings, the flap on the engine side would be set to close slightly later than that on the air intake side, thus giving the richer mixture.

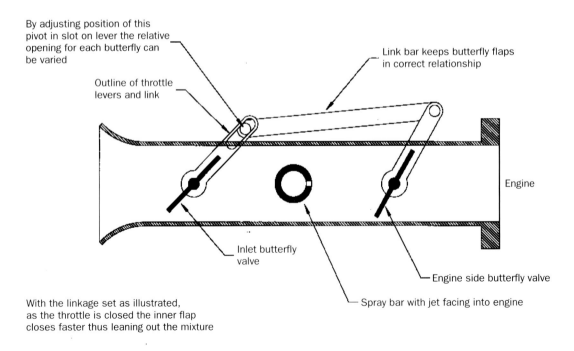

By adjusting position of this pivot in slot on lever the relative opening for each butterfly can be varied

Outline of throttle levers and link

Link bar keeps butterfly flaps in correct relationship

Engine

Inlet butterfly valve

Engine side butterfly valve

Spray bar with jet facing into engine

With the linkage set as illustrated, as the throttle is closed the inner flap closes faster thus leaning out the mixture

Layout incorporating twin butterfly carburettors, showing the adjustable linkage for varying the mixture at different throttle settings.

Triple carburettors on a six-cylinder engine; patience is needed to set this number of carburettors up correctly.

CARBURETTORS ON MULTI-CYLINDER ENGINES

Multi-cylinder engines may have more than one carburettor in order to provide the best mixture distribution to all cylinders. The problem is that this makes the setting up much trickier.

A common misconception is that the carburettor for multi-cylinder engines with a single carburettor needs to be larger than that for a single-cylinder engine of the same individual cylinder size. This is not true, however, for four-cylinder engines, because, although there are two intake strokes for every revolution, these do not occur at the same time (although there may be some overlap due to the valve timing). At any instant the air flow through the carburettor is the same as for a single-cylinder engine. The only difference is that, in the single-cylinder engine, the flow is intermittent; in the four-cylinder engine it is more or less continuous. The choke size is related to the cylinder bore and this still holds true.

Engines with more than four cylinders may need to have the size of a single carburettor increased slightly because more than one cylinder may be on the intake part of the cycle at the same time.

If more than one carburettor is used, the same rules will apply. If each carburettor serves no more than four cylinders, the size is the same as for a single such cylinder.

Care should also be taken that the fuel pipes allow for the correct amount of fuel to be delivered to each carburettor, otherwise the fuel/air proportions may not be consistent for each cylinder. Full-size engines with multiple carburettors have used balance pipes linking the different intake tracts together. These were connected at points between the carburettors and the engine and had the effect of equalizing the suction on the carburettors.

One reason for using multiple carburettors may be convenience; it is often easier and

115

The twin carburettors on the Matchless G45 (see picture, page 91) are working replicas of the originals and are supplied from a single central float chamber.

neater to use twin carburettors on engines such as flat-twins rather than having one carburettor mounted on a long curved intake manifold. This also avoids any mixture problems due to long intake tracts, which may allow the atomized fuel to coalesce into larger droplets, with possible detrimental effects on engine performance.

When making carburettors for engines with multiple carburettors, it is important that they are identical so that the settings can be replicated across all carburettors. If the carburettors behave differently, it will prove very difficult to set everything up to obtain consistent running of the engine.

It may also be a good idea to set the engine up with a single carburettor first so that it can be bedded in and tested to eliminate any other problems before attempting to set up multiple carburettors.

16   Exhaust Systems

Exhaust systems on four-stroke engines are generally very simple. Their purpose is to take the waste gases away from the engine and discharge them at a convenient point to the atmosphere. They may also provide a silencing effect. In their most basic form they are no more than a simple expansion chamber with the exhaust gases entering at one end and leaving at the other.

There are a number of refinements, for example, the creation of a labyrinth effect so that the gases are forced round a tortuous path before exiting the silencer.

Exhaust systems must be able to stand up to the heat from the exhaust gases; on petrol and glow-plug engines this means silver-soldering or welding. The easiest materials to use for exhaust systems are copper or brass tube, both of which are easily soldered, are resistant to corrosion and can be plated if desired. Stainless steel is also suitable and most grades can be silver-soldered if the correct flux is used.

Aluminium alloy is commonly used for exhaust components in model aircraft in order to save weight, but in this case assembly is often by mechanical means such as screwing or bolting together, or by using light press-fits in conjunction with a high-temperature grade of Loctite.

For engines that are to be bench run only, the exhaust system can be very simple and only has to take the burnt gases away from the engine.

Exhaust systems used in models will obviously need to be designed to fit in the

Exhaust manifold and silencer for four-cylinder engine. The silencer is constructed of silver-soldered brass and chrome-plated commercially. The exhaust manifold is welded stainless steel and the silencer is attached with a split clamp.

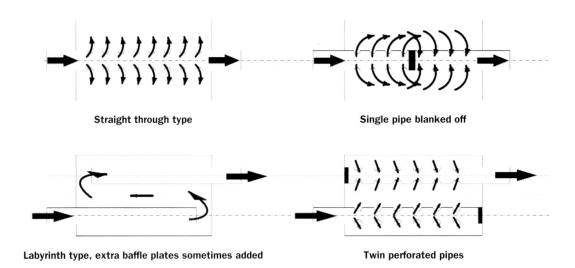

Some simple but effective exhaust systems. The straight-through type has the minimum effect on performance but silencing is not so good. The volume of the chamber should be at least six times cylinder volume; it may be filled with sound-deadening material if desired.

model, often with scale exhaust outlets and some means of trapping any oil in the exhaust.

Silencing will also be important for engines used in model boats and aircraft and in this case a good labyrinth-type silencer will be needed. For effective silencing, there must be no leaks in the system, therefore close attention must be paid to the fit of any joints.

The four-cylinder engine in the Mercer T35 Raceabout, built by Ingvar Dahlberg in Sweden. This engine has side valves driven by twin camshafts on each side of the cylinder and twin spark plugs in each cylinder.

17 Spark and Glow-Plug Ignition Systems

Spark and glow-plug ignition systems have proved to be troublesome for model-engine builders, because it is difficult to scale ignition systems down to a sensible size without sacrificing efficiency.

(Compression ignition is not covered here because it is solely dependent on the high compression ratio of the engine and the fuel mixture, which contains ether for its operation. It is also very rare in four-stroke engines, although one or two examples have been constructed.)

SPARK IGNITION

In the early days of miniature engine construction, builders had to use parts from full-size engines or wind their own ignition coils. Nowadays, however, miniature ignition components are available from several suppliers and are suitable for both mechanical points and full electronic ignition systems.

The difference between mechanical spark ignition and electronic ignition is that, in the former, the power to the coil is switched by a set of cam-operated contacts. In the latter, the switching is done by means of transistors, triggered either by mechanical contacts or electronically using a Hall Effect transistor triggered by a rotating magnet. The advantage of electronic ignition is that it is more reliable on small sizes and usually gives a much better spark. Some builders question the benefits of using electronic ignition with mechanical points, but with this method the condenser can be eliminated and the contact points are only carrying a small current, so arcing is avoided.

NB: the spark occurs when the power to the coil is switched off. There is no point timing the engine at the point of switching on and wondering why it does not run.

When the power is switched to the coil, a magnetic field is set up. This decays rapidly when the power is cut, causing a surge in the coil low-tension winding. This is amplified in the high-tension coil to produce a high-voltage surge, which arcs across the spark-plug points. The amplification is not in direct proportion to the ratio of the number of turns in the low- and high-voltage windings. The principles of such coils and magnetic induction theory are outside the scope of this book but can be found in physics textbooks or other sources such as the internet.

One of the advantages of spark ignition is that the point of ignition can be controlled precisely, by moving the contact points or Hall Effect detector in relation to the top dead centre position of the crankshaft. Setting the ignition point earlier is known as 'advancing' the ignition; setting it later is known as 'retarding' the ignition. The correct settings will depend on the speed and load on the engine, with the ignition more advanced for high-speed running and retarded for slow speeds.

Spark-ignition engines do not need special fuels to run, making them generally cheaper to operate than glow-plug or compression-ignition engines.

Mechanical Spark Ignition

This type of spark-ignition system uses a cam-operated contact breaker to switch the power to the coil and thus to provide the spark. The contact breaker consists of two tungsten contact points, arranged so that they are insulated from one another, and located so that they are opened and closed by means of a cam on the camshaft or sometimes the crankshaft.

A small condenser is connected across the points to avoid sparking at the points and also to accelerate the rate of decay of the magnetic field to produce a good spark.

The coil is connected via these points to a battery and the high-tension winding is connected to the spark plug (or distributor for multi-cylinder engines). The distributor is basically a high-voltage switch driven off the same drive as the points and connecting the correct plug to the high-tension side of the coil.

The contact breaker cam must close the points for long enough during the engine cycle to allow the coil to energize fully before the points open (break) to create the spark. If the closed period is too long, battery power will be wasted and the coil may overheat. Something between 30 and 60 degrees of closed period is sensible for most engines with modern coils and batteries. This is known as the 'dwell angle'. The dwell angle is affected by the gap between the points when they are open. Increasing the gap reduces the dwell angle while reducing the gap will increase the dwell angle. The points gap should be set around 0.020in.

The cam normally operates on a Tufnol or fibre block on the points and the surface of the

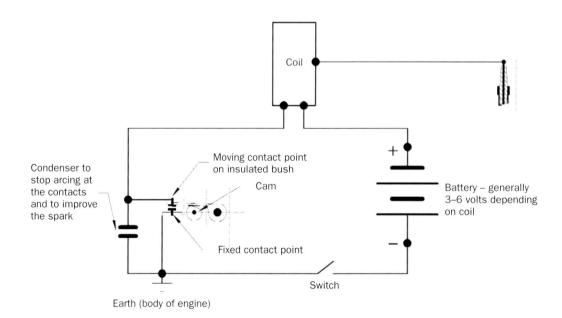

Spark-ignition circuit for use with mechanical points. Note that the spark occurs when the contacts open.

cam needs to be hardened and polished to avoid wear.

For larger engines, small car or motorcycle contact points can be adapted, but smaller engines will need to have sets of points made up. Many published engine designs provide full contact breaker details on the drawings. It is important to use tungsten contacts, which can be obtained from several sources.

It is also possible to use a micro switch for the points on slow-running engines, but it is necessary to ensure that it is rated adequately for the job (at least 10 amps) and that the 'normally open' contacts are used.

The best batteries to use are rechargeable lead acid or Nicad cells, which can be charged easily between running sessions and can provide the high currents necessary to power the coil.

A typical coil is the 'Modelectric' coil, which can operate on anything from three to six volts and is small and light enough to fit into any model. Some constructors also use small motorcycle coils, but these are a bit large for model aircraft use and may overpower small spark plugs.

The condenser needs to have a value of around 0.1mfd and to be rated at about 150 volts working. If the condenser is faulty, the spark will be very weak and accompanied by arcing at the points. The easiest way to check for a faulty condenser is to replace it.

When wiring up ignition systems the wire used must be able to handle a minimum of 5 amps, and preferably 10, and all connections should be properly soldered to avoid problems with loose connections.

The spark plug high-tension connection will need adequate insulation or clearance around the wiring to avoid the spark arcing across to earth. The most suitable wire is the flexible cable sold in electronics dealers for use with electrical multi-meters. It is very flexible and is designed for use with fairly high voltages.

Electronic Ignition

There are two types of electronic ignition. The

'Modelectric' ignition coil and capacitor.

first simply uses mechanical points to switch a small transistor circuit, which then switches the heavy current to the coil. The advantage of this is that the condenser is no longer needed and, as the current switched by the points is very small, there is no sparking. In addition, the points can be made much smaller and do not need to be made from tungsten, which makes life easier for the home builder.

The second type of electronic ignition system replaces the points by either an optical detector or, more often, by a Hall Effect transistor triggered by a magnet.

A Hall Effect transistor is simply a switch that is turned on and off by a magnetic field. This field is supplied by a very small, powerful magnet mounted on a disc on the camshaft. As the magnet passes the transistor the current is switched on. The dwell angle for such systems depends on the diameter of the magnet and its distance from the shaft centre.

Optical systems use a light-sensitive transistor that is triggered using a rotating disc with a hole in it. When light passes through the hole, the coil is switched on and the spark occurs at the point at which the transistor is covered by the disc. The dwell angle is governed by the size of the hole in the disc. One advantage of optical systems is that fibre-optic cable can be used to direct the light pulses to the transistor, making it possible to keep it away from the high voltages in the distributor on multi-cylinder engines.

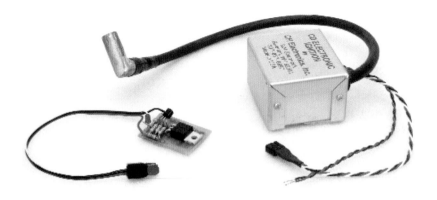

Electronic ignition systems: (left) a simple system built from a kit, and (right) a commercial system.

These systems can be made very small, and electronic circuits for Hall Effect systems can be obtained ready assembled or as a simple kit for self-assembly. Of course, those with enough knowledge of electronics could design and build their own system.

The only note of caution about these systems is that, if the connections are not set up correctly, the delicate electronic components can be damaged. Apart from that, electronic systems are ideal for model engines because they are very compact and produce a good spark (especially with some of the more complex capacitor discharge circuits). Be warned that being on the wrong end of the spark from these is not pleasant!

Two sizes of spark plug for model use, 10mm and ¼inch.

Spark Plugs

Spark plugs for model engines can be purchased or manufactured by the constructor. The size of plug will be dictated by the size of the engine. Plugs are available with the standard 10 × 1mm spark-plug thread and also ¼in by 32tpi threads, which are suitable for most engines. Note that the 10mm spark-plug thread is finer than the standard 10mm metric thread and a special tap will be needed to cut the thread in the cylinder head.

Rim fire spark plugs are supposed to be better for small engines and are reputed to be less likely to oil up in use. Kits of parts for home construction for this type of plug are available.

In the past constructors making their own plugs have used a variety of materials for the insulator, including machinable ceramic materials, cartridge fuse casings and PTFE. The ceramic materials and cartridge fuse casings are often bonded to the metal body using high-temperature epoxy adhesives. These work well because the engine temperatures in small engines are generally lower than in the full-size equivalent.

The electrodes can be made from stainless steel or mild steel, for similar reasons.

Multi-Cylinder Spark-Ignition Systems

Ignition systems for multi-cylinder engines can adopt the same principles as on full-size

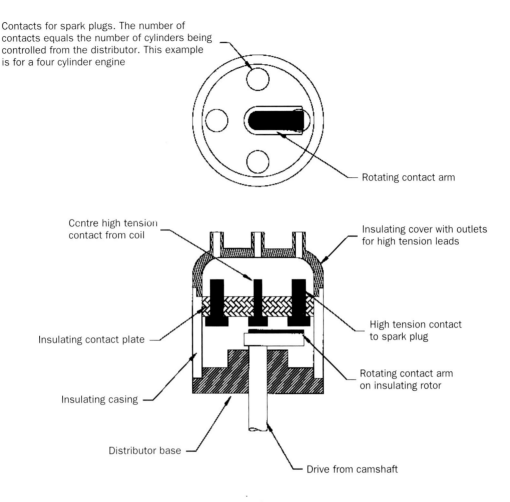

Contacts for spark plugs. The number of contacts equals the number of cylinders being controlled from the distributor. This example is for a four cylinder engine

Rotating contact arm

Centre high tension contact from coil

Insulating cover with outlets for high tension leads

Insulating contact plate

High tension contact to spark plug

Rotating contact arm on insulating rotor

Insulating casing

Distributor base

Drive from camshaft

Distributor system for four-cylinder engine. In use, the rotating contact arm connects the high-tension supply from the coil to each spark plug in the correct firing order.

engines, with one coil feeding a distributor that connects the high-tension output to each cylinder in the correct firing order.

The distributor has a rotating metal arm (the rotor arm) that connects a central electrode to each of the output leads in turn. The arm is insulated from the rest of the engine and is driven from the camshaft. The distributor is often built as a unit that includes the contact points or electronic ignition sensor. It is important to note that the rotor arm does not touch the fixed contacts. The spark jumps across the very small gap between the two.

The distributor body and rotor arm can be made from any suitable insulating material, such as Tufnol, or more modern equivalents, such as Delrin. The contacts are usually made of brass, as is the rotor arm.

Another option that can be used with electronic ignition systems is to have separate ignition systems for each cylinder, each driven from its own Hall Effect sensor, which is switched by the magnet on the camshaft. Many modern cars use such systems, avoiding the use of a distributor and providing better reliability in use. The problem with this approach for model engines is that the coils are quite large in relation to the engine and look out of place if they are not concealed from view. Cost is also a factor, with a four-cylinder engine needing four coils and four ignition circuits, equating to about £200 at today's prices.

MAGNETO IGNITION

One other option that may be considered for ignition is the use of a magneto to provide the spark. The magneto is in effect a small generator driven by the crankshaft, which generates a pulse of power at the correct point in the cycle. This feeds a coil through a set of ignition points, thus generating a spark. The coil may be incorporated into the magneto to form a completely self-contained unit.

The problem for the model constructor is obtaining the special steels used for the magnetic core of the unit.

The advantage of the magneto is that no separate power supply is needed, which makes the system attractive for a situation in which weight reduction is important. However this is less important with the use of modern Nicad batteries, which provide high amounts of power in a small size.

Miniature magnetos and components are currently available commercially.

GLOW-PLUG IGNITION

Glow-plug ignition systems are in common use in commercial miniature engines. They offer high performance and, once started, the engine does not need a power supply. The fuel used is a mixture containing methanol and possibly additives such as nitro methane to improve performance.

The glow plug looks like a small squat spark plug but instead of electrodes it has a small platinum coil that is heated by an external battery for starting. Glow plugs run on either 2 volts or 1.5 volts; care should be taken not to apply 2 volts to plugs rated at 1.5 volts. For the lower-voltage plugs 1.2-volt Nicad batteries should be used rather than non-rechargeable dry cells.

Once the engine is running, the power supply is disconnected and the coil is kept hot by the heat of combustion and the catalytic action of the methanol fuel on the element.

The point of ignition is controlled by the compression ratio of the engine, the amount of nitro methane in the fuel and the heat rating of the glow plug. Considerable experimentation will usually be needed to establish the correct fuel mixture and plug combination for a particular engine.

Glow-Plug Ratings and Types
Glow plugs are manufactured in different heat ratings to suit different performance and engine requirements. The ratings vary from 'cool' to

A selection of glow plugs: (left to right) a basic medium plug, a hot plug, a special four-stroke plug, and a plug with idle bar for better slow-running performance.

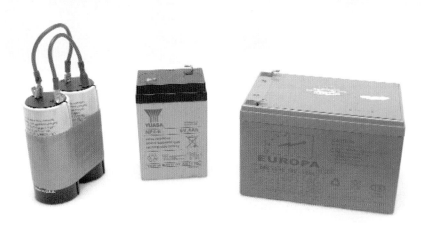

A selection of glow-plug batteries: (left) Cyclon lead acid cells (centre and right) sealed lead acid gel cells. Nicad batteries may also be used.

'hot' and will vary between manufacturers. Generally, a hotter plug will suit four-stroke engines, where the plug has to remain hot during the intake and compression strokes. There are special plugs sold for four-stroke engines, which have a shield to help retain the heat in the coil.

A hotter plug will tend to advance the ignition point and a colder plug will retard it. Adding more nitro methane to the fuel will also advance the ignition timing, and may require a cooler plug. Higher compression ratios also advance the ignition point and therefore will require a cooler plug.

Glow-Plug Power Supplies

In order to perform efficiently for starting, a glow plug must be provided with an adequate current. This means using a Nicad or lead acid battery of some sort. Suitable rechargeable self-contained glow-plug starter units can be obtained for single-cylinder engines. These clip directly on to the glow plug and provide enough power for several starts.

For multi-cylinder engines, a large power pack is needed because the current drawn by each plug may be of the order of 3 or 4amps. For example, a six-cylinder two-stroke engine will need something like 20 amps for starting. A suitable set-up might involve two 2-volt Cyclon cells in parallel.

One other point to make is that all the wiring must be substantial enough to take these power levels and connections should also be capable of handling the power. Many problems with starting glow-plug engines are down to inadequate wiring or connections.

18   Setting Up a New Engine

Setting up a new engine after assembly involves setting the valve timing correctly and also setting the ignition timing for spark-ignition engines, so that the full range of advance and retard is available and the fully retarded position is identified for easy starting.

SETTING THE VALVE TIMING

The easiest way to set the valve timing is by means of a timing disc mounted on the crankshaft, with a temporary pointer fixed to some convenient point on the engine. The timing disc does not need to be very elaborate; a simple cardboard disc will be more than adequate and a stiff piece of wire will serve quite well as a pointer.

Assuming that the cams are correctly aligned on the camshaft, the timing disc needs to be marked with a convenient valve-opening point and the top dead centre position. For a single-cylinder engine, this can be the exhaust-opening point. At this point the inlet valve is closed, and so there is no load on the camshaft from the valve springs which makes adjustment easier.

In use, the engine is rotated to the top dead centre position and the disc is tightened on to the crankshaft, with the pointer at the top dead centre mark. The crankshaft is then moved until the pointer is at the valve-opening point and the camshaft is turned until the relevant valve is at the point of opening. The camshaft drive is then fixed in this position.

One of the difficulties is in determining the point at which the valve is about to open. With pushrod engines, a dial gauge can be used on the rocker arm to detect any movement and the valve set to the point just before any detectable valve movement takes place.

With overhead-camshaft engines, where the valves act directly on the tappets, things become more difficult. One method is to set the crankshaft in the correct position and then to rotate the camshaft using the fingers, until the resistance as the cam comes up to the tappet is felt. This is the point at which to set the timing. This is more difficult with multi-cylinder engines and it may be better to use a pointer fixed to the camshaft to show the position of the cam.

Another method involves marking the midpoint between a valve closing and the point at which it starts to open on the timing disc. If the crankshaft is set so that this point is opposite the pointer, the relevant cam nose will be facing directly away from the tappet (on a flat head, pointing directly upwards), and so can be set to this position.

Once the timing is set, it pays to mark the top dead centre point on the crankshaft and also the position of a suitable part of the camshaft drive gear, relative to a mark on the camshaft housing. This allows everything to be easily reset should the engine need to be disassembled at any time.

For multi-cylinder engines, check the correct timing for all cylinders and that they fire in the correct sequence.

It is sensible not to fit permanent fixings

such as keys until you have run the engine and carried out any fine-tuning needed. It is not unknown to have to alter the valve timing after some testing!

IGNITION TIMING

The ignition timing for spark-ignition engines should be set so that the points open at top dead centre in the fully retarded position to give easy and safe starting. This can be carried out using the same timing disc as used for the valve setting.

For mechanical point ignition, a volt meter can be connected across the points through a battery to show when the points open. The contact breaker can be adjusted so that this happens at top dead centre on the firing stroke.

For electronic ignition systems the situation is more complex and, indeed, some of these systems can be damaged if they are not connected properly, with all the correct earth connections in place, and the spark plug in the circuit.

In this case, the spark plug should be mounted in a metal bracket on the cylinder head and the point of sparking should be checked as the engine is turned over to set the timing.

Some electronic systems have an LED that lights up when the coil is being energized and this can be used to set the timing to the point when the LED goes out, which corresponds to the points opening on a mechanical system. In this case, if the LED operates with the coil disconnected, the ignition timing can be set without the coil connected.

The ignition adjustment should allow for approximately 20 to 30 degrees of advance, to provide enough adjustment for high-speed running.

Once all these settings have been made, carry out a final check to ensure that the four-strokes happen in the correct sequence and that the

A timing disc used for setting the valve timing on an overhead-camshaft engine. The pointer is stiff wire clamped to the engine-mounting lug.

ignition fires at the correct top dead centre position when both valves are closed. This can be checked by observing that both the cam noses are facing away from the tappet at this point, or that the rockers are in the closed positions.

Other tuning and setting operations and fuel choices will be covered in Chapter 21.

Part III: Other Engine Types

19 Two-Stroke Engine Construction Details

The construction details of two-stroke engines differ in a number of ways from those employed for four-stroke engines. Any part of the engine not described further in this chapter should be assumed to be identical to a comparable four-stroke engine.

INTRODUCTION

The obvious difference between two-stroke engines and four-stroke engines is the lack of mechanically operated valves on the two-stroke engine. This is because control of the inlet and exhaust cycle is carried out using ports in the cylinder wall, which are covered or uncovered by the piston.

This means that a two-stroke engine is much simplified, but at the expense of some compromises to the inlet and exhaust timing of the engine. In order to overcome some of these compromises, the inlet is often controlled by a separate crankshaft-driven valve, which allows the intake period to be controlled independently of the piston movement.

INDUCTION OPTIONS

The most common types of induction system used are cylinder-wall induction, crankshaft induction, rotary-valve induction or reed- ('automatic') valve induction.

Cylinder-Wall Induction

This type of induction system was the most common in early two-stroke engines but it suffers from some disadvantages over the later systems.

In its usual form, the inlet tube is connected to a port in the cylinder wall, which is uncovered by the bottom edge of the piston on its upward stroke. The upward movement of the piston causes a reduction in pressure in the crankcase so that, when the valve opens, fuel mixture is sucked into the crankcase. As the piston descends in the next part of the stroke, the port is covered by the piston and the mixture in the crankcase is compressed, ready to be admitted to the cylinder via the transfer port.

The big disadvantage with this system is that the inlet period is symmetrical about top dead centre, meaning that the length of the induction period is limited, which restricts the performance of the engine. The advantage is its inherent simplicity.

In early engines the inlet into the crankcase was a simple drilled hole in the cylinder wall, which made this type of engine easy to construct.

Later designs used a rectangular port in the cylinder in an attempt to improve performance.

Crankshaft Induction

In order to overcome the disadvantages of cylinder port induction, an alternative

Another early engine dating from the 1930s The 10cc Ohlsson stepped fin two-stroke showing the cylinder wall induction.

Ohlsson cylinder showing clamped-on transfer and cylinder wall-induction ports with spray bar connecting directly to cylinder port.

Crankshaft from a shaft-induction engine, showing the amount of metal removed when using a plain rectangular port.

induction system was developed, using the crankshaft as the valve. The advantage is that no extra moving parts are added, although it does have the potential to weaken the crankshaft. The crankshaft induction method was, and still is, very widely used.

In its basic form, a port in the crankshaft lines up with an aperture in the crankshaft bearing as the crankshaft rotates. The width of the aperture and port controls the inlet opening period. The crankshaft port is connected to the crankcase by means of a hole bored down the centre of the shaft. The crankshaft will

Rear-disc induction parts from a high-performance two-stroke engine, showing the disc valve, shaped inlet port and drive hole for the crankpin in the disc.

normally be made larger than a plain shaft to accommodate this.

The use of such a system allows designers to specify the induction period of the engine independently of the exhaust and transfer periods, resulting in much increased performance. In order to gain the maximum opening period, and to avoid failures, crankshafts and bearings of large diameter were introduced on some racing engines. Their larger size meant that they were also heavier and for this reason, among others, the rotary inlet valve was introduced.

Rotary-Valve Induction

The rotary valve is an extension of the crankshaft-induction principle, in that the inlet timing is still controlled by crankshaft rotation, but in this case uses a separate valve driven from the crankpin. Some early commercial engines were offered with either system; those who desired higher performance at the expense of greater complexity opted for the rotary valve. Many well-known engines, including the famous *ED Racer*, used this system.

The rotary valve consists of a disc of suitable material (often Tufnol or similar), which has a cut-out in the edge. This disc is pivoted so that, as it rotates, the cut-out covers and uncovers a port in the crankcase rear. The disc is driven by an extension of the crankpin.

The advantage of this system is that the induction period can be as extended as desired without affecting the crankshaft strength. It also makes experimenting with different

An early Edgar Westbury two-stroke design, the 6cc Atom Minor *Mk III. This engine has rear disc induction. This example built by Peter Gain. Drawings and castings for this engine are still available today.*

induction periods easier because all that is needed is the manufacture and fitting of a new disc.

The rotary-drum system is a variation on rear-disc induction, which has a rotating drum with the ports cut in it driven by the crankpin. This system is a cross between crankshaft induction and rear-disc induction.

Machining Rotary Disc Valves
Machining rotary valves is easily carried out in the lathe, with the cut-out machined in the mill, ideally using a dividing head or rotary table to ensure accuracy.

The size of the intake cut-out and its relationship to the crankpin drive hole and intake port are critical for correct timing and the drive hole must be a good fit on the crankpin. Some engines have the operating edges of the cut-out chamfered to aid gas flow and, if this is done, care must be taken not to weaken the seating edge.

Because the cut-out will cause the valve to be out of balance, surplus material can be removed from the back of the disc opposite the cut-out. This is often only done with steel disc valves.

The operating face (the face running against the crankcase rear) must have a good finish, as

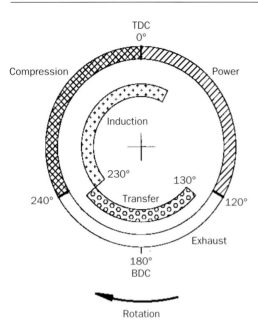

Two-stroke timing with disc induction, showing how the induction phase is longer and starts much earlier than with side-port induction.

must the corresponding crankcase face. The lubrication system of two-strokes helps in this respect because the oil in the fuel will help provide a good seal.

The disc can run on a silver-steel pin screwed or pressed into the crankcase rear, or can be provided with a shaft running in a bearing in the crankcase. In either case, correct alignment of the shaft with the disc is critical if the disc is to seat correctly and provide a good crankcase seal.

Reed-Valve Induction

The 'automatic' or reed valve, which has been widely used, consists of a thin flexible metal reed covering the intake port inside the crankcase. The reed is not driven directly but opens and closes due to the pressure variations in the crankcase caused by the piston movement.

One perceived disadvantage of this type of valve is that the opening and closing is not precisely controlled in relation to crankshaft rotation, the implication being that the inlet timing may vary. This did not stop many commercial engine builders using it. The American Cox company in particular employed the system for many years.

Like cylinder-wall induction, the reed-valve system gives symmetrical porting and means that the engine will start and run in either direction.

The reed system has also been used on multi-cylinder two-stroke engines because it helps to make construction much easier. In this situation, the reed valves are often located on the side of the crankcase to serve each cylinder, although the inlet ports may still be connected to the carburettor by a common inlet tract.

The reed valve is still in use, often constructed from more modern synthetic materials, which are more flexible and therefore result in better valve response.

The induction timing with reed valves tends to be variable because the effect of the gas flow through the port is to tend to delay the valve closing; this may be greater at high speed.

If a metal shim reed is used, care must be taken when cutting the reed out, to ensure that the edges do not have burrs or get distorted. It may be worth getting a few reeds cut commercially using water-jet or laser cutting. Many companies will carry out this work from a user-supplied CAD file quite cheaply.

It is also sensible to have some sort of open cage round the reed valve to prevent it opening too far.

PORTING

There are a number of inlet options, therefore, but what about two-stroke timing and in particular porting?

Varying the Relationship Between Exhaust and Transfer Ports

The timing diagram for a typical two-stroke engine is largely symmetrical. Moving away from the cylinder-wall induction system can allow more flexibility with regard to the

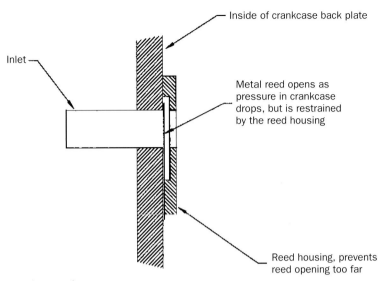

Inlet

Inside of crankcase back plate

Metal reed opens as pressure in crankcase drops, but is restrained by the reed housing

Reed housing, prevents reed opening too far

Reed-valve induction layout diagram.

induction part of the cycle. This still leaves the exhaust and transfer ports opening symmetrically about bottom dead centre and symmetrical to each other. In order to alter this situation, designers have tried several different techniques, one of which involves moving the crankshaft sideways in relation to the centre line of the cylinder. This is known as a De Saxe layout and has the effect of offsetting the angles of top and bottom dead centres. The direction of the movement is such that the connecting rod angle is reduced on the power stroke, which reduces the sideways pressure, thus improving mechanical efficiency.

The effect of the alteration of the relative angles of top and bottom dead centres is to give a more effective power stroke. There is also a beneficial effect on the transfer part of the stroke.

The other way to vary the relationship between the exhaust and transfer parts of the stroke is to have one side of the piston crown lower than the other. This obviously gives more flexibility when laying out the ports in the cylinder wall and may allow a better layout to be employed. One problem with this occurs with engines employing piston rings, because

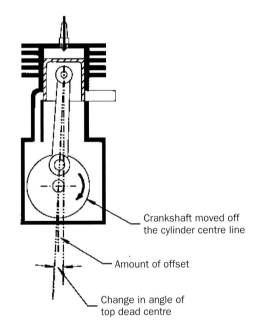

Crankshaft moved off the cylinder centre line

Amount of offset

Change in angle of top dead centre

De Saxe cylinder layout showing how the position of top dead centre (and bottom dead centre) move away from the cylinder axis. Because the BDC movement is greater than that at TDC, the transfer and exhaust timing is improved giving a better power stroke.

133

the rings obviously have to be below the lowest part of the piston crown; as a result, one set of ports may not have such a good seal as the other.

Scavenging

Another important aspect related to porting is scavenging – the use of the flow of fresh mixture into the cylinder to help remove (scavenge) the spent exhaust gases. Efficient scavenging can be accomplished in a number of ways.

Before looking at scavenging, it is useful to examine a simple engine with the exhaust ports opposite the transfer ports and a simple flat-topped piston. With the engine at the end of the power part of the stroke, the exhaust ports open, followed a fraction later by the transfer ports. The fresh mixture entering the cylinder is directed across the top of the piston and will tend to travel straight out of the exhaust ports, leaving exhaust gases at the top of the cylinder,

which will dilute the mixture for the next power stroke.

Many engines employ a deflector on the piston crown to direct the fresh mixture up towards the top of the cylinder to help push (scavenge) the exhaust gases from that part of the cylinder. This may pose a problem, since a similarly shaped cylinder head will be required, to maintain a suitable compression ratio. This will make the manufacturing process more complex.

Another approach is to angle the ports upwards as they enter the cylinder, again forcing the new mixture to the top. This makes the cutting of the ports more difficult and it may not be easy to get sufficient angle into a thin cylinder wall; as a result, many designs used thicker and heavier cylinder liners.

In order to improve on this situation, engine designers have altered the ports and provided two transfer ports on opposite sides of the

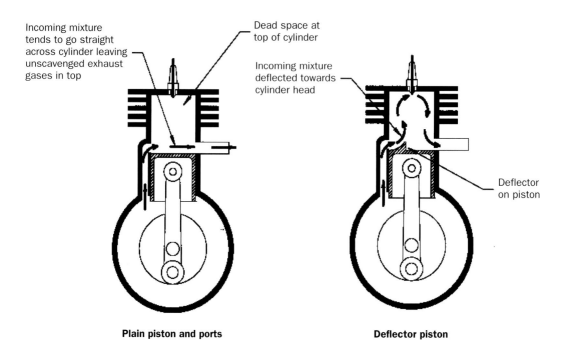

Plain piston and ports

Deflector piston

Comparison of plain and deflector pistons on engine scavenging. The deflector pushes the incoming mixture towards the head, thus helping to clear the exhaust gases from all parts of the cylinder.

A (used) deflector piston and shaped cylinder head, showing the more complex machining needed for this design.

cylinder, with the exhaust ports between them. This cylinder layout is called 'loop scavenge' because the transfer ports are angled upwards so that the new mixture is directed up towards the top and back (away from the exhaust port) of the cylinder and then loops back down towards the exhaust ports. This layout may also sometimes be referred to as Schnuerle porting.

Many other variations on this principle exist, some with up to six ports, some with the transfer ports immediately below the exhaust, but all are attempting to do the same thing. The problem for the home constructor is that some of these arrangements can be very difficult to make on home equipment, because the cylinder liner ports and crankcase passages have to line up accurately if the desired performance is to be achieved. In addition, cutting the ports may involve some hand work in the home workshop.

Cylinder liner and piston from 10cc racing engine showing very long transfer ports, with corresponding cut-out in piston skirt for ultimate performance.

Crankshaft of a 10cc engine showing the elliptical induction port, maximizing gas flow but maintaining the crankshaft strength compared with the rectangular port.

CRANKSHAFTS

Crankshafts for two-stroke engines with rotary-disc or cylinder induction are the same as for a comparable four-stroke engine. Crankshafts for engines with crankshaft induction will need to be of larger diameter than on the equivalent four-stroke engine, to allow for the cutting of the induction port and passage in the shaft.

The port is best cut with an end mill, with the passage drilled from the crank web end. Needless to say, the port must be accurately cut, even more so than for disc valves, because the smaller diameter of the shaft compared to a disc valve means that any error is magnified in its effect on the inlet timing.

If ball races are used for the shaft, a plain bearing centre section must be fitted around the induction port, to provide the necessary seal for the port operation and crankcase.

CYLINDERS

The cylinders for two-stroke engines can be made from the same materials as for four-stroke engines but, because the ports are cut in the cylinder wall, they may need to be thicker, to retain adequate strength. Cast-iron liners in particular will need attention in this area; for high-performance engines with large ports, they may not provide adequate strength unless they are cast into the crankcase.

It is more common than in four-stroke engines to find the fins turned as part of the cylinder, with no separate cylinder jacket.

Cutting Cylinder Ports

Because in the two-stroke engine the ports are cut into the cylinder itself, the cylinder is more difficult to machine. Additionally, if piston rings are used, consideration must be given to the port layout because it is possible for piston rings to catch in the ports if the latter are not suitably designed.

In early engine designs the ports were simply holes drilled in the cylinder liner, with corresponding passages machined or cast into the crankcase. The *Brown Junior* had soldered-on transfer and inlet passages with the exhaust left as a ring of holes. This type of porting is obviously easy to produce but is not very efficient and not suitable for short-stroke high-performance engines.

Later engine designs use rectangular ports, which provide greater port area for a given depth but are more difficult to cut. Of course, commercial production methods often use tools that are not commonly found in the home workshop. The best way to cut such ports is in the mill, either with a slot drill or a tee slot cutter. If the corners of the ports are to be rounded, the cutting can easily be carried out with small end mills.

Ports that consist of a series of square holes (typically where piston rings are used) can be drilled and then filed to shape with needle files. Care must be taken when filing and it is suggested that a circular clamp be made to fit round the liner as a guide when filing the edges of the ports. Another option for cutting square ports is to use a broach, but this equipment is not commonly found in amateur workshops.

Remember that a small error in cutting the position or height of a port can have a significant effect on engine timing, and hence on performance. It is also important to ensure that no burrs are left after cutting the ports.

If cylinders are made from steel that is then hardened, the ports must be cut first. In this case, the cylinder will need honing or grinding after the hardening process. Cast-iron cylinders may be lapped or lightly honed after cutting the ports to ensure that all burrs are removed.

PISTONS

Because the two-stroke engine depends on the pumping action of the piston in the crankcase to operate, the piston must form a good seal for the ports. Consequently, the piston may be considered to be a slightly more critical part on a two-stroke engine than on a four-stroke. For this reason, small two-stroke engines often have plain cast-iron pistons lapped to fit the cylinder bore.

One example engine has a composite piston, with the top third being made of cast iron, and an aluminium skirt. Obviously, the intention is to gain a good seal, and at the same time to minimize the weight penalty incurred by using a cast-iron piston.

Larger engines may use piston rings, but their use can cause problems if the ports are not designed to allow for this. The problem is that, if wide ports are used, the ends of the rings may enter the port and catch the port edges, causing severe engine damage. One way round this is to align the ends of the rings between the ports and to retain the alignment with a small pin in the ring groove, which locates between the ends of the ring.

Another method – seen on one 10cc tether car engine – incorporates a very wide ring to allow the use of large cylinder ports without the risk of a ring catching in the edges. This ring width is around one-third of the piston depth

Brown Junior *engine, showing basic exhaust and soldered transfer and inlet.*

Square-section transfer ports in a two-stroke engine cylinder liner.

and is also pinned to stop it rotating in the groove.

Another piston-related problem may occur with two-stroke engines: if the ports line up with the ends of the gudgeon pin, some method other than simple end pads may need to be employed to locate the gudgeon pin. This problem may be avoided by careful location of the ports, so that the ends of the pin are in line with a solid area of the cylinder. The fixings can take the form of small internal circlips or wire clips fitted into the ends of the gudgeon-pin holes in the piston. Another method is to fit a socket grub screw into the underside of one of the gudgeon-pin bosses inside the piston.

CRANKCASES AND CYLINDER JACKETS

Unlike four-stroke engines, the crankcase of a two-stroke engine must be sealed so that a good pumping action can be maintained, and induction and transfer can take place efficiently. Many starting problems with two-stroke engines are due to leaks in the crankcase, either through a worn bearing or badly fitted gasket.

Two-stroke engines may have separate cylinder jackets, but these are often quite short, to allow for the transfer passages, which are often machined into the crankcase wall.

One important aspect of two-stroke engine design is to ensure that the crankcase internal volume is kept low, in order to ensure a high pumping efficiency.

The volume should be such that the down stroke of the piston generates a compression ratio of around 6:1. This means that the internal space must be just enough to allow for the parts to rotate without having any excess space. This is often evident in the design of the crankcase rear cover on single-cylinder engines, which, rather than being a basically flat plate as in a four-stroke single, is often recessed into the crankcase almost up to the crankpin.

The rear cover can be fixed using a bolted flange or (often seen in smaller engines) can be screwed directly into the crankcase. In this case, provision must be made for the use of a special spanner during fitting and removal. The use of sealant on assembly must also be considered carefully because some sealants will set very hard over time and this will make a screwed fitting very difficult to remove if it gets into the threads.

The use of properly fitted O rings is the best option for screwed back plates.

Wide piston ring used on a 10cc racing engine, showing the locating pin in the groove. The ring is held in place using a ridge on the inside, which locates in a groove in the piston.

Crankcase components for built-up and one-piece designs: (left) a one-piece crankcase and the back plate; (right) the separate cylinder, crankcase and back plate for a built-up design.

Machining Crankcase Transfer Passages

The majority of two-stroke engines have the transfer passages machined into the crankcase side and these have to line up accurately with the transfer ports in the cylinder liner. One of the easiest ways to achieve this is to design the engine so that the lower edge of the cylinder jacket lines up with the top of the transfer port. This means that the transfer passage in the crankcase can then be easily machined with an end mill.

Some engines have cylinder liners with a flange round the exhaust ports, which blanks off the top of the transfer passage in a similar way when the liner is installed. This is very often the case with screwed-in cylinder liners.

The other option with engines where the crankcase and jacket are in one piece is to use a small tee slot cutter to machine the passage. Care must be taken that the top of the passage lines up with the cylinder port, and the machine indexes will be needed. It may also be necessary to extend the shank of the cutter to be able to reach far enough down the crankcase.

Those making engines from castings may find that the passages are cast in with commercial items, or will be able to provide for this when making their own castings. Cast-in passages can often benefit from cleaning up using small burrs in hand-held mini-drills, but

care must be taken not to alter any critical dimensions of the passages.

The same comments apply when machining two-stroke crankcase castings as for four-stroke engines, although it may be necessary to make extra holding fixtures to enable holding the casting to machine some of the extra ports and passages.

Crankcases may be built up to ease machining and in this case machining fixtures may well be essential. Care must also be taken to ensure correct fitting of the parts, to avoid potential leaks.

Cylinder Jackets

Cylinder jackets for two-stroke engines differ from those for four-stroke engines because the exhaust and possibly the inlet ports may be incorporated into the cylinder jacket. Engines may be designed to avoid this situation but this may not be practical or desirable in all situations.

Because of the need to provide for exhaust and inlet ports, the jacket is often machined from square-section material if castings are not used. For multi-cylinder engines with separate cylinders, this may be needed anyway.

Cylinder jackets may be attached to the crankcase using long bolts or studs that also hold the cylinder head, with a bottom flange

Cylinder jacket machined from a casting and incorporating cast-in exhaust flanges. This jacket also has a cast-in cylinder liner.

Basic turned air-cooled cylinder jacket.

and short studs or bolts, or in some (usually smaller) engines, screwed directly into the crankcase. The disadvantage of this method of fixing is that it makes lining up the ports more difficult. This method of fixing is more often found in small engines; one early commercial example is the series of *Elfin* engines from the 1940s and 50s. If screwed fixing is used, provision must be made for the use of special tools for assembling and disassembling the engine.

CYLINDER HEADS

The cylinder head on a two-stroke engine is a very simple component, with no valves or camshafts to complicate things, but the internal shape can be very important for efficient scavenging.

Compression-ignition engines, which normally have an adjustable contra piston to vary the compression ratio, will require a substantial cylinder head with the compression screw adjustment incorporated in it. This

consists of a simple screw in the centre of the head, which, when screwed in, forces the contra piston down the cylinder.

Glow-plug or spark-ignition engines will require a threaded hole with a flat seating for the spark or glow plug. This is normally in the centre of the head, although there are glow engines with twin plugs, one in the centre and another at an angle on the inlet port side.

Engines with piston deflectors may have a shaped slot in the head to correspond with this so that the piston can still rise high enough up the cylinder to achieve the necessary compression ratio. This can make machining the head difficult, necessitating the use of specially ground tools. It can be made easier by using a separate internal head, held in place by the visible part of the head that has the cooling fins on it. In effect, the visible part of the head clamps the 'real' head in place.

Construction can also be simplified by making the head in such a way that it fits into the top part of the cylinder, thus allowing any slots for the deflector to be machined right across the head. This will allow the use of specially ground fly cutters to cut a close-fitting slot for the deflector, but will also require the cylinder to be made longer than normal, to accommodate this.

When machining cylinder heads, a machining fixture will often be needed, unless the casting has a holding spigot incorporated. In the case of heads machined from bar stock, the head can be left on the parent bar until near the end of the machining process.

Heads with circular combustion chambers will need to be machined using a spherical turning tool; for a small, one-off engine, it may be possible to use a form tool ground from gauge plate and hardened.

EXHAUST SYSTEMS

Exhaust systems on two-stroke engines may be more critical than those on four-stroke engines because a badly designed silencer can have a far greater effect on performance than it would on a four-stroke. This is because gas flow restrictions and resonances in the exhaust systems of two-stroke engines can have an adverse effect, causing restriction of the gas flow out of the exhaust port.

This can be turned to advantage – if the exhaust system is arranged so that, at the time the exhaust is opened, the pressure at the port end of the pipe is reduced, then increased flow through the port will result in better clearance of the exhaust gases.

A simple length of straight pipe may well generate some improvement in performance, as will a megaphone-type exhaust.

It is possible to design a system that creates a low pressure at the exhaust port as the port opens, and then creates a reflected pressure wave that increases the cylinder pressure as the exhaust port closes. This has the effect of raising the pressure of the mixture in the cylinder, and this can give an increase in performance. Such exhaust systems are commonly known as tuned pipes and it is important to remember that the effect is dependent on the rpm of the engine, so that a tuned pipe must be designed and set up to suit the on-load speed of the engine. Assuming that the basic pipe design is suitable, this is usually done by altering the distance between the engine end of the pipe and the exhaust port. This adjustment is only for very fine-tuning, and it is no use trying to use an unsuitable pipe for the engine in the hope that it can be 'tuned in'.

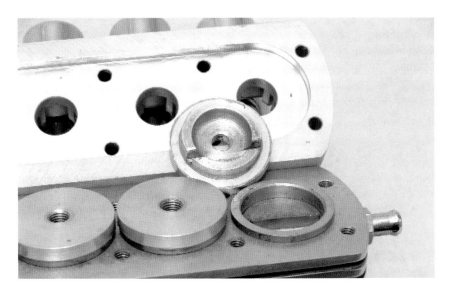

Dummy cylinder head on a six-cylinder two-stroke engine. The use of the individual round cylinder heads makes machining the complex shapes easier. All six heads are held in place by the long outer head seen at the rear.

It is important on such systems that there are no leaks between the engine and the tuned pipe. Engines designed for tuned-pipe use often have the exhaust outlet pointing to the rear with O ring seals incorporated.

For side-exhaust engines, an adaptor will have to be made up to fit on the engine exhaust flange with a connection for the tuned pipe. An O ring seal can be incorporated into this. The O rings used should be of the high-temperature silicon type.

The other aspect of tuned-pipe exhaust system design is, of course, the silencing effect. This is achieved by having an expansion chamber that allows the exhaust gases to expand before reaching the open air, as in the equivalent four-stroke system. Although tuned exhaust pipes may have a silencing effect, often an extra silencer is added to the outlet end of the pipe.

Silencers can be of the same type as used for four-stroke engines but it is not a good idea to fill them with steel wool or other damping medium, because this will soak up the oil in the fuel. For the same reason, it is sensible to provide a drain plug at a suitable point so that any collected oil can be removed.

IGNITION SYSTEMS

Ignition systems for two-stroke engines are identical to those on four-strokes except for the wide use of compression-ignition systems on small two-strokes.

The compression-ignition two-stroke engine is very commonly known – although not strictly correctly – as a 'diesel' engine. This type of ignition relies on the use of low flash point components in the fuel mixture and a high compression ratio (around 18:1), in order that the mixture can be ignited by the heat generated by the high compression of the mixture. Ether is the low flash point component in most 'diesel' fuels for miniature engines.

In the majority of compression-ignition engines, the point of ignition can be controlled by varying the compression ratio to achieve best running. This is carried out by means of a screw-adjusted contra piston at the top of the cylinder.

Some older engines often had a fixed compression ratio and relied on a very rich mixture being used for starting.

Because of the high compression ratio, the compression-ignition two-stroke engine is

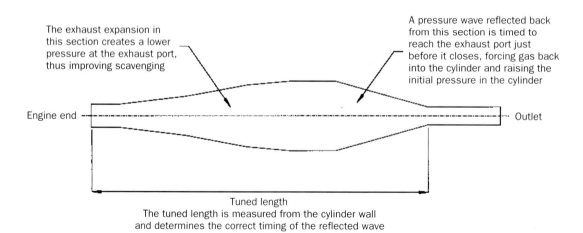

Tuned-pipe exhaust system layout. The pipe is 'tuned' to a particular RPM level and will only provide a power boost at this speed. The tuned length is usually set by trial and error with the engine in operating conditions.

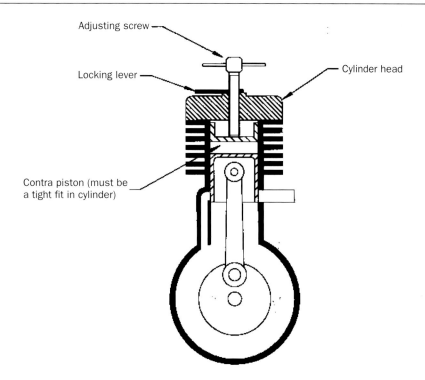

Compression-ignition cylinder head showing contra piston and screw compression adjustment. The compression is adjusted to give the best running with the minimum compression ratio.

generally more substantially built than the equivalent spark or glow-ignition engine.

The big advantage of compression-ignition engines is that they do not require any electrical power for starting or running, which reduces the amount of equipment that the operator has to carry around to boating pools or flying sites. All you need is some small tools and a bottle of fuel.

CONTRA PISTONS

The contra piston is generally made from cast iron, or may be made from steel. The advantage of steel is that it is less liable to crack in use, which can sometimes happen with cast iron.

The contra piston can be designed in two ways: it can either have parallel sides and be lapped to a tight fit in the cylinder, or it can be made with a very slight taper (one degree included) to the sides. In the latter case the sides are made quite thin so that the piston will compress slightly as it is pushed into the cylinder. In either case, a mandrel will be required for final finishing to size. In the case of parallel-sided contra pistons, a standard external hone or lap can be used to obtain the final fit.

For tapered contra pistons, a special lap can be made up with the required taper and used until the end of the piston just enters the cylinder. This will give the correct fit.

The adjusting screw is a simple screw with a cross bar at the top for adjustment. It is sensible to make the thread tight, because this will help to hold the compression adjustment and avoid the need for springs or locking bars, although many commercial engines had one or the other fitted as standard.

20   Other Engine Types

The amateur builder may wish to look at other types of engine, some of which will be described. It is not possible to provide as much detail as for the other engines in this book, but rather to introduce the builder to the type of engine and to provide some information as to the principle of operation, together with some of the issues to be considered before embarking on construction.

SLEEVE-VALVE ENGINE

In a sleeve-valve engine the port opening and closing is not just dependent on piston movement but also on the movement of an additional moving cylinder (the sleeve) between the piston and the normal fixed cylinder liner. The sleeve is driven from the crankshaft and moves up and down so that ports in the sleeve will line up with the cylinder ports, thus controlling the timing.

Sleeve-valve engines can be made as two-strokes or four-strokes, but with four-strokes, the sleeve also has a twisting motion in addition to the vertical oscillating motion. This makes the drive more complex, but allows proper control over the four-stroke cycle. Some early sleeve-valve four-stroke engines used two sleeves, one inside the other, to accomplish the same effect.

For two-stroke engines, the sleeve valve will allow more control over the timing events and can also be used in a 'uniflow' form where the fresh fuel/air mixture enters at the bottom of the cylinder and the burnt exhaust gases exit from the top. This provides two benefits: the gas flow is not subject to sudden changes in direction, and scavenging is improved.

Sleeve-valve engines were very common in full-size aircraft in the 1940s but suffered from lubrication problems between the sleeve and the cylinder, partly due to the lack of modern high-quality materials and manufacturing techniques. With model-size engines, this is less of a problem and several constructors have made miniature sleeve-valve engines, some of them models of the complex multi-cylinder aero engines of that time.

Sleeve valves also had some use in land-based applications, but were not as common in that area. They were used in both boats and road vehicles. In fact, the design patented by American Charles Knight had twin concentric sleeves and was used in cars and other vehicles for a considerable period.

Construction

Constructing a sleeve-valve engine is not very different from building any other engine. It still has all the main components and the drive to the sleeve(s) is similar in principle to other valve gear.

The obvious difference is the manufacture of the sleeves and the cutting of the ports, but this is not so different from making a normal two-stroke cylinder. However, with single sleeves for four-stroke engines, the ports will not normally be a regular shape and so will require some thought in setting up for machining.

The timing of single sleeve-valve engines is

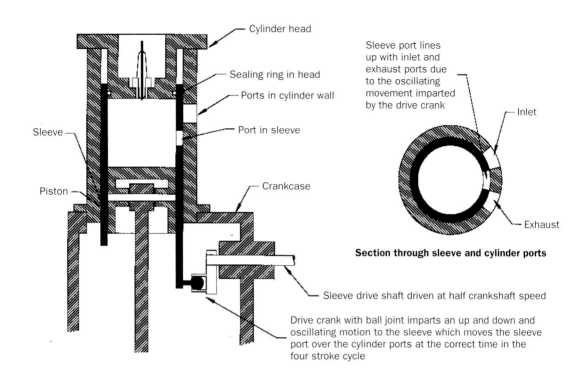

Sleeve-valve engine diagrammatic layout showing a typical sleeve-drive mechanism and also the special cylinder head with sealing ring.

complex to design because the timing depends not only on the shapes of the ports in both the sleeve and the cylinder, but also on the amount of vertical and twisting movement imparted by the drive. A CAD system is probably the easiest tool to use at the design stage. Two-stroke engines are easier in this respect than four-strokes. Remember that, with four-stroke single sleeve-valve engines, the inlet and exhaust events are controlled independently on the two opposite strokes of the sleeve, with the rotation of the sleeve causing the different sets of ports to come into play.

The final part of the drive to the sleeve will need to allow for the motion in three dimensions, so a ball joint or other form of universal joint is required.

The sleeves are probably best made from cast iron and lapped into the cylinder. Steel sleeves can be used but will require heat treatment, with possible problems due to distortion, meaning that they will need to be ground to size after heat treatment.

The cylinder head is also different because the sleeve needs to slide up on to it and must form a good seal in the process. Apart from this, the head is much easier to machine than for an overhead four-stroke engine.

One quoted advantage of the sleeve engine is that it allows more freedom to site the spark plug at the best point for ignition compared to poppet-valve engines.

Budding sleeve-valve engine constructors will find various books by the early designers such as Fedden and others, and a search on the internet will also yield a lot of information.

145

ROTARY CYLINDER-VALVE ENGINE

The rotary cylinder-valve (RCV) engine could be considered to be a variation on the sleeve-valve engine. In this type of four-stroke engine, the cylinder rotates and has a small rotary valve machined into the head. This valve opens and closes inlet and exhaust ports in the head to control the valve events and also opens the spark or glow plug to the cylinder at the correct time. The cylinder is driven by bevel gears from the crankshaft.

This principle has been around for many years and has been exploited by RCV Engines, a manufacturer of model and full-size engines.

Two variations are possible. In the first, the cylinder is extended past the head and the rotary-valve portion to form the propeller shaft. This means that the propeller runs at half engine speed, allowing the use of larger propellers in models for a given engine size. The crankshaft is just used for starting.

The second variation is more conventional, with the drive taken from the crankshaft as normal.

One advantage of the first type for model aircraft use is that the frontal area is smaller than for a conventional engine, making the engine suitable for aircraft with very stream-lined nose sections.

The timing on this type of engine is easy to design because it is similar to a drum valve, although it controls both inlet and exhaust.

Construction

Construction of rotary cylinder-valve engines should not present too much of a challenge to the home constructor, other than making the cylinder and bevel gears. This could be achieved by using built-up construction, with the cylinder bevel gear bolted on to a flange on the bottom end of the cylinder. In this case, commercial gears and a cast-iron cylinder may be used.

For those using one-piece cylinder construction, the combined cylinder and gear must be made from steel and heat-treated before being ground or lapped to final size. This latter process is more difficult than normal because the cylinder is closed at one end, so a plain lap will be needed.

In commercial designs, the cylinder and crankshaft both run in ball races, which avoid lubrication problems with the cylinder, and oil is mixed with the fuel to provide lubrication.

The unusual rotary cylinder valve (RCV) engine.

ABOVE: The RCV engine, opened up to show the cylinder and crankcase assemblies.

RIGHT: RCV engine crankcase internals, showing the connecting rod, crankshaft and cylinder drive bevel gear.

BELOW: The one-piece cylinder/prop shaft of the RCV engine, showing the drive gear and valve port (the small hole in the middle part of the shaft).

ROTARY PISTON (WANKEL) ENGINE

The rotary piston or Wankel engine (named after the inventor Felix Wankel) is very different from other forms of engine because it has no reciprocating parts. Although it follows the same inlet/compression/power/exhaust cycle as a four-stroke engine, it achieves this in its own unique way.

A specially shaped three-lobed rotor (an equilateral triangle with curved sides) rotates round an eccentric path in a shaped (epi-trochoidal) case and, as it rotates through one revolution, fuel mixture is sucked into the closed chamber formed between the rotor lobes and the crankcase bore. As rotation continues, and the closed chamber moves round the crankcase, it changes in shape, causing the mixture to be compressed and, at the point when the chamber size is at its smallest, the mixture is fired. Further rotation of the rotor allows the gases to expand (generating power) and eventually to leave via the exhaust port.

Because there are three lobes on the rotor, there are three separate chambers that go through this cycle, once for each revolution of the rotor. The engine has three full inlet/compression/power/exhaust cycles per rotor revolution, making it potentially powerful for a given size.

The inlet and exhaust ports are opened and closed by the edges of the rotor lobes, so the timing is controlled by the position of the ports in the walls of the chamber.

The rotor has an internal toothed gear, which rotates round a spur gear fixed to the inside of the chamber. This ensures that the correct relationship is maintained between the rotor and case and also allows the power to be transmitted from the rotor via the crankshaft. This is done by means of an eccentric on the crankshaft, which engages with the centre of the rotor. The crankshaft rotates three times for every complete revolution of the rotor, making the engine high-revving, although the rotor speed is one-third of the output speed. This also means that one full four-stroke cycle occurs for each revolution of the crankshaft.

On one example, the spur gear pitch circle

A Wankel rotary engine by O.S. Engines to the NSU patents.

RIGHT: Wankel engine internal layout, showing the casing internal shape, the rotor with tip seals, the eccentric shaft, and the inlet and exhaust ports into the chamber at the front.

BELOW: The Wankel cycle.

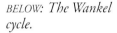

1. The rotor is rotating anti-clockwise and the face marked with the white rectangle is bridging the exhaust and inlet ports. The exhaust (right top) is about to close, while the inlet (left top) is just open. The chamber formed between the casing and the marked rotor face is at its minimum size.

2. As the rotor continues to rotate, the chamber volume increases, drawing fuel air/mixture in through the inlet.

3. The chamber volume is now at its maximum and the inlet port is just about to be closed by the top rotor tip.

4. As the rotor continues to rotate, the mixture trapped in the chamber is compressed.

5. The compressed mixture is ignited by the glow plug (bottom centre) as the rotor passes the bottom centre point.

6. The expanding mixture forces the rotor to rotate and the chamber size increases, allowing the burning mixture to expand.

7. The top rotor tip passes the exhaust port, allowing the gases to escape.

8. As the rotor continues to rotate, the chamber volume decreases, expelling the exhaust gases.

Main components of the Wankel engine: (clockwise from top left) back plate with rotor drive gear; main casing; front plate; crankshaft; and rotor, showing the internal drive gear.

diameter is the same as the diameter swept by the outer edge of the eccentric on the shaft and the gear ratio is 1.5:1.

Although the rotary piston engine has no reciprocating parts, because the crankshaft has an eccentric part and the rotor does not rotate round a circular path, the engine has to be balanced. However, because the out-of-balance forces are generated only by rotating parts, near-exact balance can be achieved by using appropriate weights on the crankshaft, making Wankel engines very smooth-running.

One major problem with constructing this type of engine is sealing the tips of the rotor lobes against the crankcase chamber. Modern machining methods and materials have helped to alleviate this difficulty.

This type of engine was first brought into commercial automotive use in the NSU RO80 sports car in the 1960s. Although it was very smooth and quiet, the engine suffered from poor reliability of the tip seals and high oil consumption. More recently, the Mazda RX8 has used a Wankel engine; the car has been very successful and the engine exhibits none of the earlier problems.

Construction

Construction of miniature Wankel engines has been achieved in the past by amateurs but is not for the faint-hearted. The greatest problem is machining the crankcase chamber to the correct shape and providing reliable rotor-tip seals.

The x and y co-ordinates for the shape of the chamber can be calculated using the following formulae:

$X = E \cos 3\varnothing + R \cos \varnothing$
$Y = E \sin 3\varnothing + R \sin \varnothing$
where
E is the eccentric throw of the rotor;
R is the distance from the centre to the tip of the rotor;
$\varnothing$ is the angle of rotation of the rotor in degrees.

The above equations can be used to produce the chamber shape as a series of XY co-ordinates, or possibly, for those with CNC facilities, to machine the chamber itself.

In miniature Wankel engines, the face seals used on full-size versions can be dispensed with by making the rotor a close fit between the ends of the casing. This means that only the tip seals need to be considered. These need to be spring-loaded in slots in the rotor tips with the outer edges (in contact with the casing) rounded. The length needs to be the same as the rotor thickness to ensure a good seal.

The rotor flanks are curved in order to increase the effective compression ratio and usually have small depressions cut into them to increase the total volume of the 'cylinder'.

If the engine is to be water-cooled, passages

will need to be incorporated into the casing for water circulation. In models, these could be simple drillings with connecting passages machined into the end plates.

All the ancillary items such as carburettors and exhaust systems are the same as for any four-stroke engine.

Wankel engines offer a lot of scope for the inventive constructor and the result will be a most unusual model.

RADIAL AND ROTARY ENGINES

Radial and rotary engines are among the most impressive of engines and will be familiar to many because of the part they played in early aircraft development. Many miniature radial and rotary engines have been built and miniature radial engines have also been produced by several commercial manufacturers. Both types have the cylinders arranged in a circle (radially) around the crankcase, with the connecting rods effectively joined at one central crankpin. After

that, the differences between the two types start to become apparent.

Both types are commonly four-stroke engines, although there have been some two-stroke radial engines produced. In both types, the fuel/air mixture is normally taken in via the crankcase and distributed to the individual cylinders. In the rotary engine, the mixture is taken in via the fixed crankshaft centre at the rear of the engine; in the radial engine, the carburettor can be fixed to the crankcase rear at a suitable point.

This means that, for both types, as with two-stroke engines, the crankcase must be sealed for the engine to run correctly.

Distribution of the mixture to the cylinders in both radial and rotary engines is usually via pipes linking each inlet port to the crankcase. However, in some early rotary engine designs the mixture entered the cylinder through a valve in the crown of the piston; this meant that only the exhaust valve was needed in the cylinder head and only one set of valve operating gear was required, as with the Gnome rotary engine.

A good example of a small five-cylinder radial glow-plug engine by Seidel at a UK model rally. This example has exposed rocker arms and has been fitted with an exhaust collector ring.

A nine-cylinder Gnome rotary engine running at a UK model show.

Rotary Engine Specifics

The rotary engine was used in many early aircraft in order to gain higher power outputs. Such engines have their own set of challenges because of the fact that the whole engine is rotating. For example, the ignition system has to take the high voltage for the spark from a coil outside the rotating engine, then distribute it to each spark plug in the correct sequence. This is accomplished by making the distributor in the form of a ring of insulated contacts round the crankshaft rear so that, as the engine rotates, each spark plug is connected to the coil at the correct point in the cycle.

The other issue is that of engine balance. Obviously, if the component parts for each cylinder are not identical, they will be of different weights when assembled and vibration will result.

The valves are generally operated by pushrods and, unlike many ordinary four-stroke engines, some means must be provided to ensure that the pushrods are secured against the centrifugal forces generated when the engine is running.

Because the whole engine rotates, the fixed crankshaft will need to be more substantial than for the equivalent radial engine.

Lubrication of the rotary engine is normally accomplished by mixing oil with the fuel, as in a two-stroke engine; although it may be pumped into the intake separately, the result is still a total-loss system. Rotary engines do throw out a lot of oil when running – spectators are advised to stand well back!

One factor worthy of note is that miniature rotary engines normally have spark ignition. This is because it would be difficult, if not impossible, to arrange for power to be applied to glow plugs when the cylinders are rotating.

Radial Engine Specifics

Radial engines are more like a 'normal' engine, the obvious difference being that the cylinders are arranged in a circle round the crankcase. In many other respects, the components will be very similar to those of any other engine. As with other types, radial engines can be made with spark or glow-plug ignition.

Unlike rotary engines, radial engines can

A fine model Gnome rotary engine seen at an exhibition. This shows the single exhaust valve in the cylinder head, which is a feature of this design. The mixture enters the cylinder via a valve in the piston crown.

A sixteen-cylinder double-row radial engine, the Bristol Hydra, *built by Brian Perkins. Only two full-size versions were built.*

The cylinder head detail on the Hydra, *showing the valve-drive arrangement. This engine was built without castings; all machining was carried out by hand.*

have a proper pumped lubrication system with an oil tank outside the crankcase, and with the oil being pumped round the engine, then collected in a small 'dry sump' for return to the tank by a scavenge pump.

A proper exhaust system can also be fitted to a radial engine, so it will be much cleaner in operation than a rotary engine.

Peculiarities Common to Both Radial and Rotary Engines

Both types of engine have a number of peculiarities and present certain design and machining issues. It is worth stressing that, for both types, some of the components are not regular shapes and, because such engines often have more cylinders (the minimum is three) than most, the use of suitable jigs and fixtures is strongly recommended.

Cylinder Layout

In both engines, the cylinders are arranged radially around the crankcase and can be in one or more rows. The reason for using more than one row is to gain more power, while still keeping the frontal area as small as possible. Additionally, the maximum practical number of cylinders in any one row is nine. All the cylinders in a row act on one crankpin, and it is not practical to accommodate the nine connecting rods without the big end becoming very large.

Both types of engine normally have an odd number of cylinders in each row to minimize vibration when running. This means that any radial or rotary engine with a single row of cylinders will have an odd number such as three, five, seven or nine. There was one full-size engine built with an even number of cylinders in each row – the 16-cylinder double-row Bristol *Hydra* – but only two were ever built because they were found to suffer from severe vibration problems when running.

The majority of rotary engines were single-row designs, although there has been an example of a double-row model engine.

Full-size radial engines were built with up to three rows of cylinders, giving twenty-seven cylinders.

Crankcase

One of the most noticeable visible components of radial and rotary engines is the large circular crankcase. Although roughly circular in shape, it will have a series of equally spaced flat faces round the edge for the individual cylinder seatings.

The size of the crankcase needs to be taken into account when choosing which model engine to build, because the machining of this item requires the use of a rotary table or dividing head that is capable of supporting the crankcase. Additionally, the milling facilities need to be capable of boring and drilling operations (for cylinder seats, and so on), with the crankcase rotating about the horizontal axis under the milling head.

One other consideration is the number of cylinders to choose. The easiest number for a first project is one that provides angles between the cylinders of whole numbers of degrees. This means that three- and nine-cylinder engines are easiest, with angles of 120 and 40 degrees respectively, followed by five cylinders with an angle of 72 degrees. This is important if you do not have a dividing head and have to use a rotary table. If you doubt this, then work out the angle between the cylinders of a seven-cylinder engine!

Most model engine crankcases of this type are machined from a suitable billet of aluminium alloy.

The crankcase front and rear may be made from castings. The crankcase front on a rotary engine must have provision for bolting on the propeller, so it must be strong enough for this, and is often turned from steel.

Crankshaft

The basic crankshaft is a single crankpin design, much like a normal single-cylinder four-stroke with rear valve drive. However, it can be quite complex, particularly in rotary engines because the fuel mixture enters through the rear of the shaft; also, such engines often have complex bearing assemblies to take the engine thrust.

Because the shaft has to drive the valve gear, and because the most common big-end layout used does not allow for split bearings, the shaft must be made in two parts in order to assemble the engine. The parts are normally bolted together with tapers and keys or splines to maintain correct alignment.

Rotary engine shafts do not need balance weights but the shafts for radial engines must be balanced.

Connecting Rods

The connecting rods in these engines are a composite item, with one 'master' and several 'slave' connecting rods (one for each of the other cylinders), and are arranged to take the drive from each cylinder.

Several different layouts for the big end have been used in the development of such engines.

The most common, and easiest to make, has one master connecting rod running on the crankpin, and each of the other cylinders having a slave connecting rod with its 'big end' running on a pin in the master rod big end.

This layout suffers from the fact that, as the engine rotates, the master connecting rod is tilted. This results in the crank pins for the other connecting rods being slightly out of line with the cylinder centre line when the main crank pin is on the top or bottom dead centre for the cylinder in question. This slightly alters the valve and ignition timing, and the compression ratios of the secondary cylinders. In practice, it is not possible to compensate totally for this alignment error because, although moving the big ends will compensate for the error at top dead centre, this will result in an increased error at bottom dead centre.

Some builders alter the length of the slave connecting rods to compensate for the compression ratio errors. The ignition timing

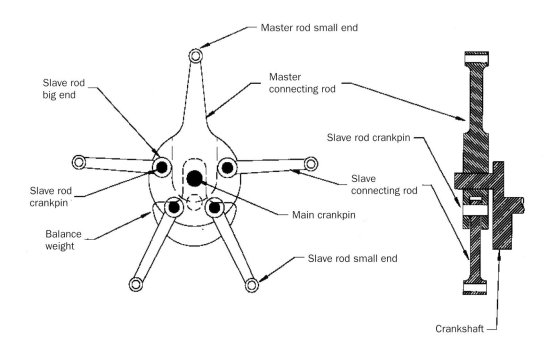

Layout of master/slave big-end assembly, showing slave rods grouped around enlarged big end on the master rod.

errors can be corrected by altering the ignition cam and distributor, so that ignition for each cylinder takes place at the correct point.

Another design of big end is known as the Manley-Baltzer, after the inventor. In this design, each slave connecting rod has a curved foot at the lower end, which is a close fit in circular grooves in the two opposite halves of the master big end. Each connecting rod is free to rotate about the main crankpin but cannot move horizontally. This layout aims to overcome the small errors in the top dead centre positions of the slave connecting rods, which are inherent in the geometry of the first design.

This type of big end is much more difficult to make than the other type, although models have been made incorporating this design.

Valve Gear

The valve gear for radial and rotary four-stroke engines is most often pushrod-operated. The components follow the same pattern as for a standard four-stroke, with the exception that rotary engines will often incorporate some form of positive retention mechanism for the pushrods. The method of driving the valves, however, is often very different, particularly with rotary engines.

The most common layout seen uses a large double cam ring driven by an internal gear from the crankshaft, and operating directly on the tappets in the front or rear gear casing. One cam ring drives the exhaust valves and the other is used for the inlet.

The cam ring can be driven in the direction

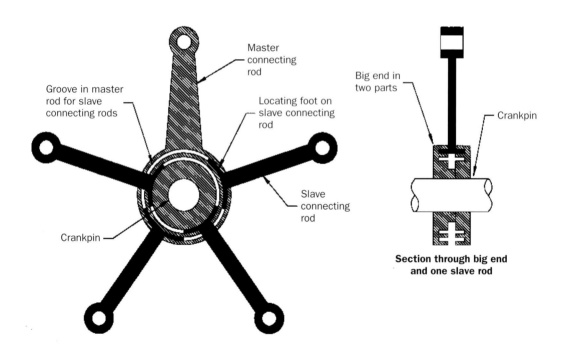

Diagrammatic view of the Manley Baltzer big-end arrangement, showing the circular grooves in the master connecting-rod big end, which hold the curved feet of the slave connecting rods. The feet can move around the grooves, allowing for the tilt of the master rod but the slaves connecting the rod centre and the crankpin always remain in line.

156

of rotation of the engine, or in reverse. The advantage of using the reverse drive is that the cam speed past the tappets is less, with the likely result that wear will be reduced.

The normal firing order for radial and rotary engines is alternate cylinders in turn; for a nine-cylinder engine, the firing order would be 1, 3, 5, 7, 9, 2, 4, 6, 8. This means that the cam must move two cylinders between one ignition point and the next. For a nine-cylinder engine, this is a convenient 80 degrees because the angle between adjacent cylinders is 40 degrees. Other numbers of cylinders are not so convenient.

There are two formulae that can be used to work out the gear ratio and the number of cam lobes required for engines (although this will be specified for published designs). Of course, it is beyond the scope of this book to cover fully all aspects of radial and rotary engine design.

The formulae to use are:

$L = (C \text{ or} - 1)/2$ (where C is the number of cylinders and L is the number of cam lobes)
and
$R = C \text{ or} - 1$ (where C is the number of cylinders and R is the cam gear drive ratio)

In both cases the + sign is used when the cams rotate in the same direction as the crankshaft and the – sign is used when the cams rotate in the opposite direction.

Taking as an example a nine-cylinder engine with a cam rotating in the opposite direction to the crankshaft, these formulae give a gear ratio of 8:1 and four lobes on each cam ring.

The cam profiles are chosen to give the correct opening periods for the valves and, because of the size of the cam rings, the tappets will be ball-nose tappets or roller cam followers.

There are also some engine designs that use individual spur gear drives operating separate camshafts for each cylinder. The problem with this is that it can be tricky to fit all the gears in the crankcase, particularly for a nine-cylinder engine.

Gear housing for a nine-cylinder radial engine, showing the gear drive to the cam gear. (Example constructed by Mike Tull.)

Sleeve-valve engines will use geared drives. They can become extremely complex, but for those building models of full-size engines they may well be prototypically correct.

Radial and rotary engines are impressive and provide some unique challenges for model engineers in their construction but, for those who have sufficient experience, a radial or rotary engine can provide a very satisfying project.

There are other engine types suitable for the home constructor and a search of patents and early books on engineering will provide much inspiration. Some examples include the Atkinson cycle engine, in which the four-stroke cycle is accomplished within one revolution of the crankshaft; the eight-stroke engine that fires once every eight strokes (for economy); and, for two-stroke enthusiasts, several versions of the split-single two-stroke, in which the function of the second cylinder is to pump the fuel/air mixture into the power cylinder, thus providing a degree of supercharging.

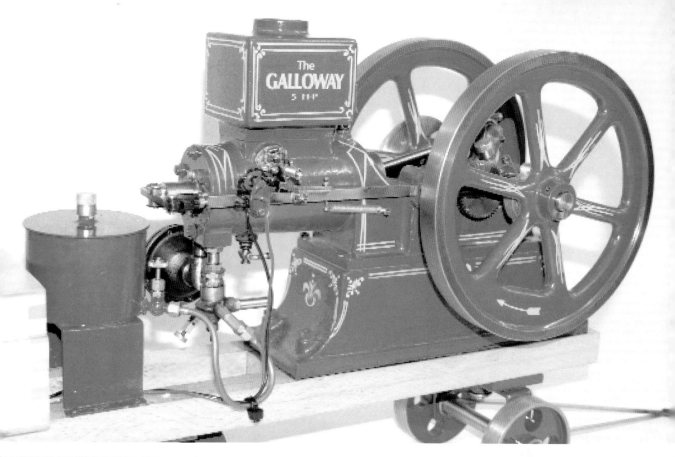

21 Engine Operation

SAFETY CONSIDERATIONS

This advice applies particularly to those running engines with airscrews because of the obvious potential for accidents. Running model engines should not be a risky operation, provided some common sense is exercised and everyone is aware of the possible dangers.

Engine Mounting

First, all engines must be mounted securely on a properly designed engine mount, which is firmly fixed down. Obviously with water-cooled engines there is no propeller thrust to contend with, but an engine coming loose while running can still cause problems.

Stationary engines are often mounted on a base complete with all ancillary items and will be designed to run on that.

It is a good idea to mount aero engines on a proper mount designed to be fixed in a portable DIY vice such as the 'Workmate'. With this type of set-up, you will need to provide some means to prevent the engine coming out even

OPPOSITE, TOP: A finely lined-out model of a Galloway 'round-rod' gas engine, built by Alan Thatcher. This model features low-tension ignition in which the contact points are actually inside the cylinder. The model is self contained and runs on its own scale trolley.

OPPOSITE, BOTTOM: Crankshaft and big-end detail on the Galloway engine.

if the vice screws should loosen with the vibration. Some metal bars bolted to the underside of the mount will prevent the mount from sliding forward.

It is also important to make sure that the vice is securely tied down to something firm. When running engines in open grass areas, you could use a device like a giant corkscrew, sold for tethering dogs for use on caravan and camping sites. This is screwed into the ground and the vice is tied to it. This will prevent the engine being pulled forward when running.

On no account, should engines be mounted in a vice by clamping them directly – not only because this may damage the engine but also because, if the vice loosens, there is nothing to prevent the engine from flying out.

Restricting Access

When running engines within close proximity to observers, it is desirable to fence off the running area to prevent spectators getting too close to the operating engine. Remember that rotating propellers are almost invisible and people will forget how large they are.

Similarly, it is sensible to install a bar, longer than the propeller diameter, immediately behind the engine as a reminder to the operator to stay well clear of the propeller.

Ideally, all spectators should be kept behind the engine.

Airscrew Security

For larger engines (above 10cc) it is advisable to provide some means of ensuring that the

An engine mounted on a running mount in a DIY portable vice, showing how to apply the electric starter.

airscrew cannot come off the shaft even if the propeller nut comes loose (as with a backfire, for example). This can take the form of a wire ring or E-clip in a suitable groove in the shaft. Many scale model engines will have a bolted-through propeller mount, which provides the same security. Commercial engines will often have special propeller nuts that lock in place.

Ignition and Glow-Plug Batteries

Modern rechargeable Nicad or sealed lead acid batteries can put out very high currents for a short time, so, if a short circuit occurs when the battery is in the tool box, a serious fire can result. Also, if a battery is short-circuited close to the operator, the resulting flash may cause serious eye damage.

Batteries should never be kept loose in a tool box and ideally should have covers to prevent accidental access to the terminals.

For the same reason, all leads and connections should be properly insulated to minimize the possibility of short circuits.

Fuel storage

Fuel of all types must be kept in properly marked, sealed containers, and should be kept away from any likely sources of ignition. When transferring fuel, use a proper funnel or a specially made fuel pump, and avoid spillage. If spillage does occur, it should be cleaned up immediately and certainly before attempting to start any engine.

LEFT: A commercial Seidel radial engine running at a show. The long bar clamped behind the engine is a 'chicken stick', to prevent the operator inadvertently walking into the spinning propeller.

Commercial fuels for glow engines will be provided in suitable containers and should be kept in those.

Accidents occur when the unexpected happens. If spilt fuel is ignited when an engine starts, that may be the occasion when a hand gets into the propeller.

Airscrews

Airscrews for model engines are available in a wide range of sizes and materials and must be treated carefully to avoid problems. They should always be balanced using a propeller balancer to avoid vibration. This can be done by removing material from the heavy end tip. This applies even to commercial propellers.

If any damage occurs to a propeller, such as chipped edges or the appearance of cracks, that propeller should be scrapped immediately. Do not forget that the propeller may well be rotating at speeds in excess of 18,000rpm on a two-stroke engine; if it fails, the consequences could be disastrous. Many commercial propellers have the maximum rotating speed specified and on no account should this speed be exceeded.

Cooling

Water-cooled engines without a built-in radiator or cooling system will need to be provided with a cooling supply when being run. This can be achieved by using two tanks connected to the cylinder block, one slightly above the engine, and one underneath, so that the water in the top tank can flow through the engine and be collected in the lower tank. Remember that the feed from the top tank must go to the bottom of the cylinder block and the outlet come from the top.

Another option is to provide an engine-driven water pump (either gear or centrifugal) and to circulate the water through the engine to and from one tank. If an extended run is carried out, resulting in the water heating up too much, cold water can be introduced during the run to cool things down again.

Air-cooled engines running on airscrews will

Airscrew safety clip, which prevents the propeller nut from coming off if loosened by a backfire, or when starting using an electric starter.

be cooled by the air flow but, if they are run without an airscrew, then a small electric blower can be positioned to blow air across the cylinder fins. It does not require much air flow to keep most engines cool.

Airscrew and Water-Screw Sizes

For both marine and aero engines, the propeller or water-screw size must be such that the engine is running at its designed rpm when the model is in motion. Commercial engines will give recommended sizes and many published designs will do likewise. If you are designing your own engine, look through the recommendations for similar commercial or other designs for some guidelines.

In order to get the best fit, you will need to test different sizes and types and to measure the rpm for each. Once the best fit has been established on the bench, flight or water trials can be carried out to see which gives the best performance in practice. The choice will be dependent on the type of boat or aircraft being used.

The rpm of engines running on airscrews

can be measured using one of the cheap optical rev counters (tachometer) that are available commercially.

Fuels and Fuel Tanks

Fuels for model engines fall into three categories depending on the type of engine: spark ignition, glow or diesel. Each has its own distinct characteristics and requirements for satisfactory operation. It is advisable to filter all fuel before use and ensure that it is kept in clean containers.

Spark-Ignition Fuels

By far the most common fuel for spark-ignition engines is straight petrol (gasoline), as used in full-size cars. This can be used straight from the pump in most engines and requires no other additives. The octane rating seems to have no noticeable effect on model engines, but those with very high compression ratio high-performance engines may find some benefit with higher-grade fuels.

Very small engines may need something more volatile and some builders use old-fashioned cigarette lighter fuel rather than pump petrol. The problem is getting hold of this, in these days of gas lighters. One substitute is the special fuel used for some small liquid-fuel camping stoves, which can be obtained from caravan and camping suppliers.

Another fuel sometimes used with spark ignition is methanol, which has been used in the past in the belief that it gives improved performance. In practice, it will only give this benefit if the compression ratio is increased.

It may surprise many builders to learn that the specific heat energy released by burning a given volume of any of the fuels at its correct fuel/air ratio is approximately the same. This means that, when using methanol, the needle valve will need to be opened considerably from the setting used for petrol, because of the lower calorific value compared with petrol, and the fact that the air/fuel ratio by weight for complete combustion is less than half of that for petrol. In one engine run on both petrol and methanol, the needle valve setting increased from 1.5 turns open on petrol to 2.5 turns with methanol.

Half-size model of a twin-cylinder Matchless G45 *racing engine, showing the electric cooling fan used when running.*

Two-stroke engines will need to have oil mixed with the fuel for lubrication. One of the proprietary two-stroke oils will be suitable for this. Castor oil can be used, but may form sticky residues in some engines. The petrol/oil ratio should be no more than about fifteen per cent for most engines and may be set lower once the engine has been run in.

A small amount of oil – about five per cent – may be added to four-stroke fuels to aid top-end lubrication. If the four-stroke engine relies on oil blowing past the piston for lubrication, the amount may need to be increased slightly.

Glow-Ignition Fuels

Modern glow fuels are methanol-based, usually with nitro methane as an additive to improve running.

Unlike petrol engines, most glow engines need a specific formulation of fuel to achieve maximum performance. For published designs, a recommended fuel mix and glow-plug combination may be suggested. If no information is available, start with a nitro methane content of around ten per cent used with a medium-grade glow plug.

For glow engines needing oil in the fuel for lubrication, synthetic or castor oils must be used, because mineral oils do not mix well with methanol. It is best to buy ready-mixed commercial fuel, which is available in many different formulations to suit most engine types. For those mixing their own fuels, the constituents can be obtained through the model trade or possibly from industrial suppliers.

It is possible to run glow-plug engines on a petrol-based fuel, but they will need to be started on a methanol mixture, and then switched to the other fuel, which may also need to contain some methanol to keep the glow plug alight.

Diesel (Compression-Ignition) Fuels

The fuel for compression-ignition engines is paraffin-based, with ether added, so that the mixture can be fired by the cylinder compression when starting. The basic fuel mixture for a two-stroke diesel is equal parts of paraffin, ether and oil. A small percentage of amyl nitrate may be added to aid smooth running.

Higher oil content is normal with compression-ignition engines.

Commercial diesel fuel is available. Users can mix their own, but it may be difficult to obtain ether and amyl nitrate for 'home' use in these safety-conscious times.

Power Supplies

Ignition Power Supply

The best source of power for spark or glow-plug ignition is one of the various types of rechargeable cells now available. There are three main types of cell in use, standard or sealed lead acid, Cyclon-type lead acid and Nicad cells. It is important to note that the first two have a voltage per cell of two volts, while Nicads have a cell voltage of 1.2.

It is also important to mention that, if damage is to be avoided, all these types of batteries must be charged with the correct charger. Car chargers are not suitable for anything other than ordinary lead acid batteries. Suitable chargers can usually be obtained from the battery supplier.

Glow-Plug Power Supplies

For glow-plug use, the plug voltage must be checked carefully. If using 1.5-volt plugs with lead acid cells, then long (2m) leads between battery and plug should reduce the voltage enough to avoid burning out the plug.

Special power supply units for flight boxes are also available. These allow different voltages to be selected and the current of the glow plug to be monitored, and are the best option for regular use. They are usually powered from a 12-volt car battery and incorporate power supplies for fuel pumps and starters.

The plug element should glow bright red if connected outside the engine.

All connections must be properly made and

A commercial optical tachometer suitable for two-, three- or four-bladed propellers.

proper glow connectors should be used at the plug end, both to ensure a good steady glow and also to avoid the possibility of short circuits. Crocodile clips are not a good idea, as they can work loose and short out on the engine.

For multi-cylinder engines, it is best to wire all plugs to a convenient connection point so that just one lead needs to be connected. The plugs will be wired in parallel, so connecting more plugs does not mean a higher voltage, but requires a higher-capacity battery. Battery capacity is measured in ampere hours; a 4 ampere hour battery will provide 4 amps for one hour or 8 amps for thirty minutes.

This is important because the glow plugs on a four-cylinder engine will draw between 15 and 20 amps when connected, so a 4 ampere hour battery will only provide about twelve minutes' starting time before needing recharge. With a reliable, easy-starting engine this is not a problem, but if the plugs become too wet or the engine proves difficult to start for some other reason, the available power will soon be used up.

It is also important to provide battery wiring of adequate thickness (30-amp capacity) and to ensure that the connections are correctly rated and firmly tightened. A loose connection will very quickly heat up with 20 amps going through it.

Spark-Ignition Power Supplies
Batteries for spark-ignition systems will depend on the ignition system and coil being used and will generally be 3, 4, 6 or 12 volts. Because the ignition system is on all the time the engine is running, large-capacity cells are a must for a full day's running. Any of the cell types can be used but the easiest is probably the sealed lead acid type, which is available in a range of voltages together with small charger units.

The same comments about wiring apply when using spark ignition although with multi-cylinder engines the electrical current required will be no greater than for a single because there is generally only one coil to energize, even though some builders are adopting modern automotive practice and using a separate small coil for each cylinder. However, the battery capacity needed will be greater because the coil is energized for longer during each engine revolution than for a single-cylinder engine, so a larger-capacity battery pack is advisable.

Fuel Tanks for Bench Running
The purpose of the fuel tank is to provide a steady and reliable supply of fuel to the carburettor. For bench running a metal or suitable plastic tank can be mounted on the running stand at a convenient point. For permanent installation, such as for a model stationary engine, a metal (brass or stainless steel) tank would normally be used, with copper pipes connected by proper screwed unions.

For other installations, the plastic tank has advantages because the fuel level is visible through the sides. A silicon rubber or plastic fuel pipe would be used. A filter should be fitted in the fuel line and also a fuel shut-off valve, particularly for diesel and glow engines. For multiple carburettor set-ups, small plastic connectors and tee pieces as used for fish-tank air supplies can make the plumbing easier.

When using a plastic tank it is important to check that the tank is suitable for the fuel being used – some plastics will soften when used with glow or diesel fuels.

It is usual to fit the tank with the top of the tank level with the needle valve, but other builders mount the tank above the needle level to provide a positive flow of fuel. A barrel-type carburettor will easily suck up the fuel from slightly below the needle and is advantageous because, when the engine stops, the fuel will cease to flow.

Float chamber-type carburettors will need a positive feed and, unless an engine-driven fuel pump is used, the tank must be above the float chamber level.

STARTING ENGINES

Some engines will be fitted with recoil starters, spring starters or, these days, even built-in electric starters, and for starting these, no extra equipment is needed.

Manual Starting

For marine engines with a grooved flywheel, a cord can be looped under the flywheel in the groove, with one end held in each hand. By pulling up on one end and lowering the other, the engine can be pulled over easily. The technique is soon mastered and several turns can be given to the flywheel using this technique.

For aero engines, the method is to flick the propeller over top dead centre with the fingers. In order to do this it is best if the propeller is set horizontally as the engine comes up on the compression stroke. This enables the engine to be smartly flicked over top dead centre and, hopefully, to start. The use of a finger protector is to be recommended.

Electric Starters

The safest and best method of starting engines these days is to use one of the battery-powered electric starters that are widely available. Such starters run from a 12-volt supply and a sealed lead acid or car battery is the ideal power source.

For marine use, a rubber belt is fitted under the flywheel, and the groove on the starter pulley looped into this before pressing the switch. Another option is to have a rubber ring on the starter pulley and to press this against the flywheel edge. This method is most often used for model cars and may be built into the running stand, so that the flywheel can be easily pressed down on to the starter to turn the engine over.

For airscrew use, the starter is fitted with a rubber cone, which can be pressed against the spinner or propeller nut to provide drive to the engine. It is important to press the starter cone firmly against the spinner before pressing the switch.

Whenever using one of these starters, it is essential first to turn the engine over by hand, to ensure that there is no hydraulic lock or other problem, as the starter will apply a considerable amount of torque and can potentially damage the engine.

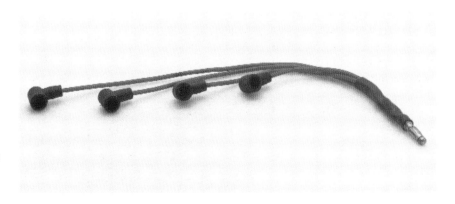

Multi-cylinder glow-plug connection loom for a four-cylinder engine.

RUNNING THE ENGINE

Once the engine is properly mounted with cooling and fuel supply, you need to follow a regular procedure for running it. Each builder will develop his or her own sequence, but the following will hopefully act as a guide for those new to the situation:

1. Check that everything is securely fixed down, and that the propeller or flywheel nut is tight.
2. Check that the fuel supply is connected correctly, the ignition connections correctly applied, the cooling system filled with water and, if applicable, the sump filled with oil.
3. Turn the engine over gently by hand with the ignition/glow and fuel off, to check that everything is free. If you flick it over at this stage, you must act as if it will start; diesel engines have been known to spring into life with the residual fuel in the crankcase in this situation!
4. With the ignition or glow-plug supply still switched off, open the needle valve to the required amount (usually between two and three turns), then, with the throttle wide

open, place a finger over the intake and turn the engine a couple of revolutions by hand to suck some fuel into the engine. This will be indicated by fuel coming up the pipe (if transparent) or by fuel on the finger used to block the intake.
5. Close the throttle to about one-third open, connect the ignition or glow plug, and either flick the engine smartly or use the starter. The engine should fire and, hopefully, run.
6. For a glow engine, allow the engine to run for 20 or 30 seconds before disconnecting the power supply.
7. Assuming it runs, open the throttle gradually and adjust the needle valve to get smooth running. The needle should be set to give the highest speed on full throttle and then opened up a fraction for best running. The throttle can now be closed gently and the best slow-running position found.
8. Once the engine is set up correctly (*see* chapter 18) for the full procedure for setting the carburettor), a slightly rich mixture may be beneficial for starting.

Diesel engines follow a slightly different procedure because of the adjustable compression

An electric starter showing the driver (on left) suitable for use with airscrew spinners and also flywheels using the V groove and a rubber belt.

ratio. When starting a diesel, it is primed as before and the compression adjustment backed off approximately half a turn. With the needle valve set slightly rich, turn the engine over smartly. If it does not show signs of life, increase the compression slightly and try again until the engine runs.

Now both the compression and fuel needle controls must be adjusted for best running. Lean out the needle valve and the engine will probably start to run roughly, so increase the compression slightly to correct this. Repeat this sequence to get best running. It is important to note that the compression setting should be the minimum for clean running. Over-compression may cause damage or premature wear to the engine.

RUNNING IN

All new engines will benefit from a period of running (breaking) in, which helps the various moving parts to bed into each other in a controlled way, and will ensure a long life for the engine.

The basic technique is to run the engine for short periods at reduced power using a rich mixture, which will help to keep the engine cool until it loosens up. With diesels, the compression can be backed off slightly as well.

During running in, the engine speed should be varied using the needle valve or throttle so that the engine is run through a range of speeds.

If the engine shows any signs of overheating and tightening up (slowing) then it must be stopped and allowed to cool down before further running.

Initial runs should be for three or four minutes, allowing the engine to cool down between each. During each run, longer periods of higher-speed running can be tried provided all seems to be all right.

The amount of running-in time needed will vary from engine to engine, but generally engines with plain bearings and plain pistons (no rings) will need the longest. Larger engines are likely to need more running in than smaller engines. For commercial engines, the manufacturer will specify the running-in regime to use.

After each run, check all the engine bolts for tightness and make sure that the engine is generally OK.

CARBURETTOR SETTING

Setting the carburettor for the best idle (slow-running) speed and smooth transition from idle to full throttle is not difficult as long as a set procedure is followed.

For a plain spray bar with no throttle, the needle valve is gradually closed until the engine is running evenly at maximum revs. Glow engines should be richened a fraction from this very lean position for best running and long glow-plug life.

For engines with a proper carburettor and throttle, the sequence will depend on the type of carburettor used. Some types have a throttle stop and a separate slow-running adjustment, while others just have a throttle stop. The slow-running adjustment may take the form of an air-bleed control or a separate slow-running needle valve that comes into operation when the throttle is closed.

For carburettors with just a throttle stop, allow the engine to warm up fully and then close the throttle gently. Adjust the stop to obtain the best stable slow-running position. Now open the throttle smoothly (but avoid snapping it open) and the engine should speed up smoothly to full power. Extended periods of slow running may result in the engine hesitating before the revs rise. If the fuel line is squeezed when the throttle is closed, the engine may speed up slightly, or start to cut out. If it speeds up slightly, the slow-running mixture is probably too rich; raise the slow-running setting slightly using the throttle stop adjustment.

For engines with a separate slow-running mixture control, set the throttle stop for the slowest even running, then adjust the slow-running adjustment for maximum speed at this

A 40cc four-cylinder double overhead-camshaft glow engine running on the bench.

setting. This cycle may be repeated to get the best setting. Again, the engine speed should rise smoothly and without hesitation when the throttle is opened.

With all carburettors, it can be a mistake to try to set the idle speed too low. This will result in unreliable throttle control.

The correct techniques will come with experience and the best way for beginners to progress is to join a local club where members are usually only too pleased to encourage and help those new to the hobby.

AFTER THE RUN

After running any engine, it should not just be put away until next time, because, particularly for glow engines with nitro methane fuels, the products of combustion are corrosive.

With any engine it is best to ensure that all the fuel is run out, by stopping the engine using the fuel tap or by closing the needle valve. This should burn any fuel out of the engine. Next, for spark-ignition or glow engines, remove the plug and put several drops of thin oil into the

cylinder, then turn the engine over by hand a few times to spread this around. The same can be done for the intake on two-strokes, to get oil into the crankcase.

For four-strokes with oil in the sump, the oil should be drained and the sump left empty.

Special 'after-run' oils are sold by the model trade and these can help prevent corrosion when an engine is not being used.

The cooling system on marine or stationary engines should be completely drained. The outside of the engine should be cleaned and wiped over with an oily cloth to remove any dust and dirt that may get into the engine. Any residue in the silencer should be drained, to prevent it running back into the engine.

A paper plug can be inserted into the intake to prevent any dirt entering during storage – make sure it is removed before trying to start the engine next time.

The ignition or glow batteries should be disconnected and put on charge ready for the next run.

The engine should be given a final check over before it is put away.

Appendix: Engine Troubleshooting

This guide should help identify potential problems with starting and running engines. It is not intended to be exhaustive but should give some pointers for most problems that arise.

Symptom	Possible Causes
Engine refuses to turn over	Engine seized due to corrosion or fuel residue Hydraulic lock on diesels
Engine will not start and shows no signs of life	Ignition/glow battery flat Glow plug open circuit Spark plug short-circuited to earth Ignition system faulty No fuel reaching carburettor Compression backed off too far on diesel engines Mechanical points dirty or incorrectly adjusted Air leaks into crankcase on two-strokes Engine flooded with fuel
Engine fires briefly but then dies	Fuel blockage or needle valve set incorrectly Loose connection on ignition system Ignition battery has very low charge Compression set too low on diesels Sticking valve on four-strokes Spark plug oiling up Glow plug (just) blown
Engine runs for a short time before slowing and stopping	Needle valve set too lean Valve sticking after getting hot Valve-clearance adjustment moving Piston or ring seizing when hot Plain bearing tightening up when hot Ignition battery discharged Spark plug oiling up Compression screw working loose on diesel

Symptom	Possible Causes
Engine runs unevenly possibly with backfire	Needle valve set too rich Ignition points sticking Ignition points bouncing (if at high revs) Loose ignition connection Ignition too far advanced Spark plug or lead shorting to earth Glow-plug grade too hot Needs more nitro methane in fuel Valve springs too weak
Engine stops when glow-plug power disconnected	Glow-plug grade too cold Too much nitro methane in fuel Needle valve set too rich
Engine stops suddenly with loss of compression	Glow or spark plug loose Head gasket leaking Valve stuck open Piston failed
Engine 'knocks' when running	Compression set too high on diesels Ignition too far advanced Glow-plug grade too hot Compression ratio too high
Engine sluggish, does not rev	Ignition capacitor blown Ignition retarded

Glossary

Barrel carburettor A type of carburettor in which the air flow is controlled by a rotating barrel with a hole that forms part of the inlet tract.

Big end The end of the connecting rod that encircles the crankpin.

Bore The hollow part of the cylinder in which the piston slides.

Bottom dead centre (BDC) The point during the rotation of the engine when the piston is at its lowest point in the cylinder.

CAD (computer-aided design) system Software used to produce engineering drawings on the computer.

Cam ring The rotating ring in a radial or rotary engine, which operates the valves.

Cam A rotating part designed to impart a reciprocal or variable motion to the valves

Camshaft A rotating shaft with one or more cams attached, normally driven from the crankshaft.

Capacity (total) The total of the swept volume plus the head volume of an engine.

Carburettor A device that mixes the fuel and air in the correct proportions for combustion.

Chucking piece A piece of metal on a casting, which is used to hold the casting for machining but does not form part of the finished part.

Compression ignition A method of ignition in which the heat generated by compression of the fuel/air mixture causes ignition to occur.

Compression ratio The ratio of the capacity divided by the head volume.

Connecting rod The rod between the piston and the crankpin, which transmits the force of the power stroke to the crankshaft.

Contact breaker The mechanical switch, driven by the crankshaft or camshaft, which switches the current to the ignition coil.

Contra piston The tight-fitting piston at the top of the cylinder in compression-ignition engines, which is used to alter the compression ratio.

Crank web The part that connects the crankpin to the crankshaft.

Crankcase The part of the engine casing enclosing the crankshaft.

Crankpin The pin that connects the connecting rod to the crankshaft.

Crankshaft The main shaft of an engine, driven by the crank(s) and transmitting the power outside the engine.

Cubic capacity *See* Capacity

Cylinder The parallel tube in which the piston moves.

Cylinder head The part that closes the end of the cylinder above the piston.

Cylinder liner The thin cylinder that forms the working surface for the piston inside the cylinder jacket.

Diesel engine A type of engine in which the heat generated by the compression of the fuel/air mixture causes ignition.

Displacement A measure of engine size. *See* 'Swept volume'.

Distributor The device that directs the high-tension voltage to the correct cylinder in a multi-cylinder engine.

Electronic ignition An electronic switching system used to switch the current to the coil.

Firing order The sequence in which the cylinders fire in multi-cylinder engines.

Four-stroke An engine in which the complete cycle occurs during four strokes of the piston: a power stroke, an exhaust stroke, an inlet stroke, and a compression stroke.

Gear pump A type of rotary pump consisting of two gears meshing in an enclosed chamber.

Glow plug A small electrically heated element that is used to trigger ignition in glow-plug engines.

Glow-plug ignition A type of ignition using a glow plug.

Gudgeon pin The pin that holds the piston and the small end of the connecting-rod together.

Hairpin valve springs A type of valve spring in which the coil is subjected to a twisting motion by two arms that connect to the valve gear.

Hall Effect transistor An electronic device that acts like a switch when subjected to a magnetic field. Used to trigger electronic ignition systems.

Head volume The volume left at the top of the cylinder when the piston is at top dead centre.

Ignition The action of starting the combustion of the fuel/air mixture in the cylinder of the engine.

Inlet tract The passage between the carburettor and and the inlet port.

Lean mixture A fuel/air mixture in which there is insufficient fuel to use all the air for combustion.

Needle valve A valve in which a pointed needle is moved in or out of a small hole. Used to alter the amount of fuel entering the engine inlet.

Oil pump An engine-driven pump used to circulate oil for lubrication.

Over square An engine in which the stroke of the engine is less than the cylinder bore.

Overhead valve A type of engine in which the inlet and exhaust valves are located in the cylinder head.

Piston The close-fitting part connected to the small end, which compresses the mixture in the cylinder and transmits the power to the connecting rod.

Piston crown The top surface of the piston.

Piston skirt The thin cylindrical lower part of the piston below the gudgeon pin.

Piston ring Thin springy rings fitted in grooves near the crown of the piston, which provide a seal against the compression of the engine and prevent oil from the crankcase reaching the combustion space.

Plunger pump A type of oil pump in which the oil is pumped using a reciprocating plunger acting in a cylinder.

Poppet valve A valve with a mushroom-shaped head, which is lifted from its seat rather than hinged.

Pushrod A rod used to transmit the reciprocating motion from the tappets to the rocker arms.

Radial A type of engine in which the cylinders are arranged in a circle around the crankcase.

Rich mixture A fuel/air mixture that contains too much fuel to allow complete combustion.

Rocker arms Pivoted arms that transmit the motion of the cams or pushrods to the valves.

Rotary An engine in which the cylinders are located radially around the crankcase and in which the whole crankcase/cylinder assembly rotates about the crankshaft, which is fixed.

Scavenging The process of clearing the products of combustion from all parts of the cylinder when the exhaust port is open.

Side valve A type of engine in which the valves are located in a chamber at the side of the cylinder below the cylinder head.

Skirt *See* Piston skirt.

Sleeve valve A type of engine in which the events are controlled by valves consisting of oscillating and rotating sleeves with ports cut into them.

Small end The end of the connecting rod connected to the piston.

Spark ignition An ignition system in which the combustion process is triggered by a high-voltage spark in the cylinder, generated by a high-tension coil.

Spark plug A device with electrodes across which the spark is created in the cylinder in spark-ignition systems.

Spray bar A tube with a small hole set in such a position that the air passing through the carburettor will atomize the fuel fed into the tube.

Square An engine in which the cylinder bore is equal to the stroke.

Stoichiometric ratio The ratio of air to fuel necessary to achieve complete combustion. Usually expressed in terms of weight.

Stroke The distance travelled by the piston from bottom dead centre to top dead centre.

Sump The closed chamber at the bottom of the engine in which the lubricating oil is held.

Swept volume The volume created by the movement of the piston from top dead centre to bottom dead centre in the cylinder. *See also* 'Displacement'.

Tappet A device used to transmit the motion of the cam to the pushrod, rocker arm or valve.

Throttle The device used to control the speed of an engine, usually part of the carburettor.

Two-stroke An engine in which the complete engine cycle takes place within two strokes of the piston.

Under-square An engine with the stroke larger than the cylinder bore.

Valve guide The cylindrical part in which the valve stem slides in four-stroke engines.

Valve stem The thin cylindrical part of the valve, which locates the valve in the guide and transmits the motion from the rocker or tappet.

Wet liner A cylinder liner that is in direct contact with the cooling medium in a liquid-cooled engine.

Index